Mastering Enterprise Solution Modeling

A Guide to Agile Enterprise Solution Architecture (A-ESA)

Sean (Chunhong) Gu

Mastering Enterprise Solution Modeling: A Guide to Agile Enterprise Solution Architecture (A-ESA)

Sean (Chunhong) Gu
Portland, OR, USA

ISBN-13 (pbk): 979-8-8688-0991-0   ISBN-13 (electronic): 979-8-8688-0992-7
https://doi.org/10.1007/979-8-8688-0992-7

Managing Director, Apress Media LLC: Welmoed Spahr
Acquisitions Editor: Aditee Mirashi
Coordinating Editor: Kripa Joseph
Copy Editor: Kezia Endsley

Cover designed by eStudioCalamar

Cover image designed by Unsplash

Distributed to the book trade worldwide by Springer Science+Business Media New York, 233 Spring Street, 6th Floor, New York, NY 10013. Phone 1-800-SPRINGER, fax (201) 348-4505, e-mail orders-ny@springer-sbm.com, or visit www.springeronline.com. Apress Media, LLC is a California LLC and the sole member (owner) is Springer Science + Business Media Finance Inc (SSBM Finance Inc). SSBM Finance Inc is a **Delaware** corporation.

For information on translations, please e-mail booktranslations@springernature.com; for reprint, paperback, or audio rights, please e-mail bookpermissions@springernature.com.

Apress titles may be purchased in bulk for academic, corporate, or promotional use. eBook versions and licenses are also available for most titles. For more information, reference our Print and eBook Bulk Sales web page at http://www.apress.com/bulk-sales.

Any source code or other supplementary material referenced by the author in this book can be found here: https://www.apress.com/gp/services/source-code.

If disposing of this product, please recycle the paper

Table of Contents

About the Author xi

About the Technical Reviewer xiii

Acknowledgments xv

Introduction xvii

Part I: A-ESA Specification and Demo 1

Chapter 1: A Brief Overview of A-ESA 3

The Intent of A-ESA 3

The A-ESA Model Specification 5

    Architecture Area 6

    Architecture View 7

    Architecture Element 11

    Architecture Property 16

A-ESA Lean Model Mode 17

Summary 21

Chapter 2: A-ESA Modeling by Example 23

A-ESA Profile 1: Strategic Architecture 27

    ESA Principles 27

    Capability Views 32

A-ESA Profile 2: Lean Business Architecture 34

    Pattern Views 34

    Use Case Model Views 35

Business Process Views 37

Process View Mapping with Value Stream 40

Architecture Outline View: Service Layering.................................42

High-level Functional Service Views 43

Functional Service Interface Views 44

A-ESA Profile 3: Reference Architecture 45

Application Business-centric Reference Architecture 46

Data-centric Reference Architecture..47

Technology-centric Reference Architecture 48

Context-based Reference Architecture...49

Baseline Reference Architecture 50

A-ESA Profile 4: Enterprise Integration Architecture........................52

Capability Views 53

Case Scenario Views 56

Outline Views...60

Metrics Views 62

Pattern Views 64

Functional Service Views 66

Deployment Views 71

A-ESA Profile 5: Application Architecture...72

Case Scenario Views 72

Architecture Outline Views 73

Metrics Views 75

Pattern Views 82

Functional Views 85

Validation Views 106

A-ESA Profile 6: Microservice Architecture (MSA) 109

    Microservice Design Views... 109

    Metrics Views 115

    Data Collaboration Views.. 120

    Operational Environment Views... 123

A-ESA Profile 7: Cloud Hosted Architecture (FaaS) 125

    Capability Views 127

    Case Scenario View 128

    Metrics View 130

    Enterprise DevOps View 133

    Functional Collaboration View 137

    Operational Infrastructure View... 138

A-ESA Profile 8: Big Data Technical Architecture 140

    Capability Views 140

    Case Scenario Views 141

    Solution Architecture Overview 142

    Operational View.. 148

Model Thinking Questions.. 152

    Questions on Profile 1: Strategic Architecture.................................... 153

    Questions on Profile 2: Lean Business Architecture 154

    Questions on Profile 3: Reference Architecture 156

    Questions on Profile 4: Integration Architecture 157

    Questions on Profile 5: Application Architecture................................. 159

    Questions on Profile 6: Microservice Architecture.............................. 161

    Questions on Profile 7: Cloud Architecture 163

    Questions on Profile 8: Big Data Technical Architecture..................... 165

Summary.. 166

Part II: A-ESA Governing Ideas 169

Chapter 3: The A-ESA Thinking Framework 171

The Distinct Nature of A-ESA Thinking... 172

The A-ESA Architectural Process 173

Pillars of the A-ESA Thinking Framework 177

Abstraction for Landscape View Architecture.. 178

System Aspect-orientation for Structural Architecture............................... 180

Metrics Mapping for Decisional Architecture 184

A-ESA Modeling Effects 185

Architectural Clarity... 185

Architectural Responsibility... 190

Architectural Quality.. 191

Summary.. 194

Part III: A-ESA Essentials.. 195

Chapter 4: A-ESA Measurement 197

Enterprise Input Metrics... 202

Enterprise Principle Guidance 202

Enterprise Governance Guidance 203

Key Architectural Quality Metrics... 204

Architectural Structuring Quality 205

Architectural Connectivity 207

Functional Quality Metrics 208

Business Functional Quality 209

Application Service Usability... 209

Application Service Adaptability 212

Application Service Maintainability 212

Operational Quality Metrics...214

    User-aware Operational Quality...214

    Non-runtime Operational Quality..224

DevOps Metrics...229

    SRE Practice 232

Solution Constraint and Opposing Force...233

    Solution Constraint Metrics 234

    Cost as a First-Class Metric 235

    Opposing Force Metrics...236

Lean Measurement Cluster 237

    S-MAPS Measurement Cluster 237

    S-MAPS NFRs Checklist 239

    S-MAPS NFRs by Team Responsibility..242

Summary...243

Chapter 5: A-ESA Modeling Styles.....................................245

Style 1: Strategic Architecture 253

Style 2: Enterprise Architecture 255

Style 3: Heat-Mapping Capability Architecture 258

    Enterprise Capability 258

    Mapping with Value Stream..263

Style 4: Business Architecture 266

    Relationship Between Business Model and Business Architecture............266

    Organization Communication Structure 270

    Business Objectives 273

    Business Scenario..274

TABLE OF CONTENTS

Style 5: Information Architecture 276

Style 6: Application Architecture 279

    Functional Architecture 279

    DevOps Environment 280

Style 7: Data Architecture 281

    Data Transaction Considerations 281

    AI Considerations in Data Architecture 284

Style 8: Technical Architecture 287

    Operational Architecture 287

    Technical Platform Architecture 288

    Physical Architecture 292

Style 9: Service-oriented Architecture (SOA) 293

Style 10: Microservice Architecture (MSA) 298

    Microservice Architecture Principle 298

    Microservice Architectural Process 299

    Microservices Architecture Design 300

    Microservice DevOps 303

    Microservice Pattern 305

    Microservice Operational Architecture 307

Style 11: EDM Architecture 308

Style 12: Large-scale Website Architecture 312

Style 13: Agile and Digital Architecture 321

    Agile Architecture 321

    Digital Architecture 330

    Agile and Digital Architecture 331

Style 14: Cloud Architecture 332

    Cloud Hosting Architecture 332

    Cloud Architecture Migration 333

Cloud Service Selection..336

Style 15: Security Architecture..338

Style 16: Reference Architecture...342

   AI Reference Architecture and Pattern 345

Architectural Evolution..349

Summary..351

Chapter 6: A-ESA Governance Techniques.....................353

Guide 1: Customizing Modeling for Purpose 355

   Adopting A-ESA Model Form(s) 355

   Following a Modeling Process..357

   Customizing A-ESA Architectural Style...............................358

   Customizing A-ESA Solution Operating Model...................358

   Customizing A-ESA Areas/Views 359

   Customizing A-ESA Elements 363

Guide 2: Applying Leading Practice..367

   Architectural Tenet..367

   Architectural Pattern and Template 369

Guide 3: Key Tradeoff Consideration 372

   Similarity and Dissimilarity Analyses...................................372

   Key Architectural Considerations..377

   Comparative Rating Techniques 381

Guide 4: Application Architecture Techniques 384

   Serviced-based Application Technique..................................384

   Example of Application Partitioning Practices.....................393

   Application Domain Structuring (Stages) 398

   Microservice Domain Structuring (Steps)............................400

   Application Data Service Structuring 402

Guide 5: Technical Architecture Techniques...........................404

Deployment Package Mapping 404

NFR-driven Operational Excellence 405

Technical Data Service Structuring 407

Guide 6: Effective Governance 412

Governance Authority 412

Governance Enforcement 413

Architecture Validation 415

Governance Automation 416

Guide 7: Solution Management Context 418

Architectural Assessment 421

Requirements Management 423

Cost Control and Management 430

Test Management 434

Risk Management 436

Asset Management 437

Guide 8: Architecture Adoption Event (AAE) 439

Summary 442

Epilogue 443

Appendix I: A-ESA Spec Addendum 445

Appendix II: A-ESA Tools 461

Appendix III: Modeling Language Comparison 463

Appendix IV: ESA Architect's Roles and Skills 465

Appendix V: FAQs about A-ESA 471

Bibliography 485

Index 491

About the Author

 **Sean (Chunhong) Gu** is a current Open Group architect certification board member and a certified *distinguished architect* with broad IT experience in various industries, including international business, manufacturing, healthcare, and retail. Over the past 30 years, he has led a multitude of large-scale enterprise solution projects in the United States and China, and has recently been focusing on enterprise solution architecture consulting and training. His previous book, *Service-oriented Enterprise Application Architecture*, drew from his numerous landing projects and sold very well in China.

As a global IT architecture SME (Subject Matter Expert) and a chief IT architect at IBM GCG, Sean facilitated thousands of architects and technical leaders in China, the Asia-Pacific, the United States, and Europe, and he mentored top-level executive architects and IT solution leaders in different industries.

Sean graduated from the University of Chicago and holds masters' degrees in physical science, systems analysis, and business administration. He has a wide array of interests, including swimming, hiking, calligraphy, and the zither. He now resides in Los Angeles, California, United States.

About the Technical Reviewer

 **Jordan Wang** is a proven and competent information technology professional with excellent technical, organizational, and communication skills. He has 20 years of IT experience in the FSI, telecom, and energy industries within a multi-language and multi-culture environment. Eighteen of those years was in cloud consulting, designing, building, and operations. He is familiar with GenAI technology, LLM, NLP, and CV, including Open AI services and ChatGPT, IBM WatsonX, Azure Machine Learning, and Google Bard. He is familiar with architecture design methodologies and frameworks, including traditional IT, the hybrid cloud architecture, and the main AI technologies.

Acknowledgments

I want to express my gratitude to the publishing team, especially Aditee Mirashi and Kripa Joseph at Apress and Springer, for helping me bring this book to life. I also want to acknowledge Jordan Wang for his technical review of the book. I am deeply appreciative of the encouragement and support I have received from my colleagues, business partners, customers, students, and fellow IT architects in writing this book.

Introduction

Solution modeling embodies the unity of knowing and doing.

This book serves as a guide to applying the Agile Enterprise Solution Architecture (hereafter referred to as Agile ESA or A-ESA),[1] which provides a link between Enterprise Architecture (EA) and Solution Architecture (SA), leaning toward the latter in an Agile manner. The book covers the architectural thinking and measurement process for better understanding and pragmatic adoption. It includes various architectural styles that use A-ESA to produce simple and comprehensive insights.

Enterprise architects often create IT plans, blueprints, or processes that are not grounded in reality. IT professionals often rely on codebases, reengineering, or design models to explain how a system works. However, these alone may not be sufficient to manage complex solutions or to explain why or how a solution system was directed, valued, built, hosted, evolved, and most particularly, architected.

The book extends the Agile ESA specification and presents an architectural philosophy with six constituents: the thinking framework, method specification, measurement metrics, governance technique, applicable situation, and tool support. These six constituents form a well-rounded A-ESA model, as shown in Figure 0-1. Each constituent is covered in detail, as outlined in Table 0-1.

---

[1] See "Agile Enterprise Solution Architecture – An IT Service-based Modeling Approach" for detailed specifications.

"Talk is cheap" is a motto that was often quoted by my Silicon Valley team during my earlier IT career. The book takes a practical approach and explores how the enterprise solution architecture is embodied through a systematic philosophy and these six constituents, accompanied with examples. It begins with a brief introduction to A-ESA and eight typical solution cases (Part I), followed by a thinking framework (Part II). Part III then explores the essentials of A-ESA modeling in more depth.

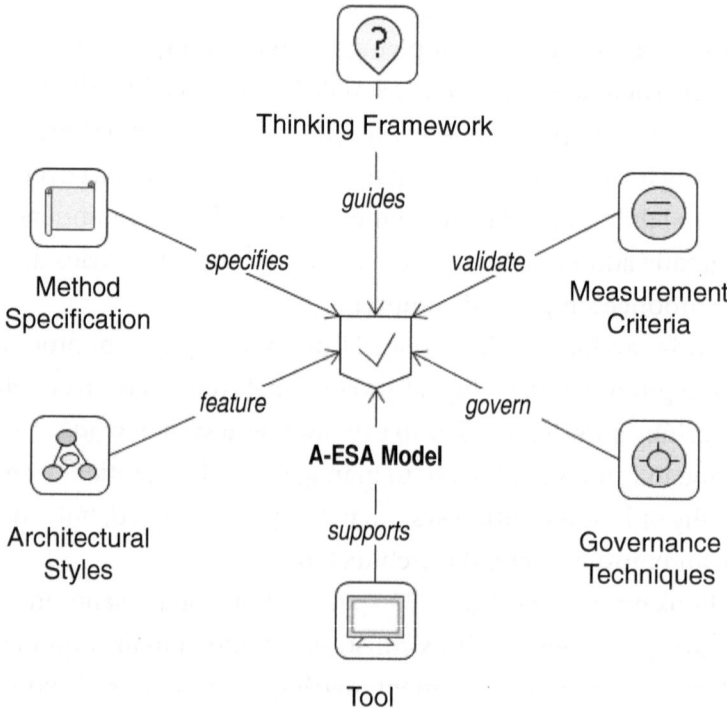

Figure 0-1. *The essence of the Agile ESA model*

Table 0-1. *A-ESA Constituents*

Constituent	Part	Chapter	Description
Model specification	I	Chapter 1	Method specification: the A-ESA foundation
		Chapter 2	Examples of A-ESA modeling
Thinking framework	II	Chapter 3	Architectural thinking: governing ideas in A-ESA
Measurement	III	Chapter 4	Measurement metrics by core architecture area
Architectural style		Chapter 5	A-ESA modeling for distinctive styles
Governance		Chapter 6	A-ESA governance techniques
Modeling support		Appendix II	A-ESA tools

PART I

A-ESA Specification and Demo

CHAPTER 1

A Brief Overview of A-ESA

Build a simple, significant, and systematic model to attain solution architectural viability.

This chapter summarizes the A-ESA modeling approach as a prerequisite for the chapters that follow.

The Intent of A-ESA

Agile Enterprise Solution Architecture (A-ESA) aims to provide a *link* between Enterprise Architecture (EA) and Solution Architecture (SA), leaning toward the latter, which covers both software and system architectures. Agile refers to a more practical and less rigid approach. A-ESA acts as an intermediary between various enterprise-wide considerations and solution concerns, including strategic directions, business needs, data analysis, application design, and operational engineering.

To wit, A-ESA covers a little bit of strategic business architecture, some enterprise architecture, some general solution architecture, and a little bit of solution design. A-ESA refers to a large and complex enterprise solution

© Sean (Chunhong) Gu 2024
S. Gu, *Mastering Enterprise Solution Modeling*,
https://doi.org/10.1007/979-8-8688-0992-7_1

or set of solutions, NOT a relatively independent project solution, a simple solution, a software application, or a solution design that is not governed by EA. When a solution is small, short-run, or obvious and everything fits in your head, you don't need an ESA or even an SA. On the other hand, the problem with a large-scale architecture is that it tends to over-architect, so A-ESA provides simplicity and flexibility when dealing with significant case scenarios and intuitive IT service-based[1] cross-cutting views.

The A-ESA model is also called the *S3 model*. This is because it aims to achieve these three "S" (*simple, significant,* and *systematic*) objectives:

- **Simple** specifications for easy learning and adoption:

  - Specify a less formal and minimal set of model elements at the just-enough, coarse-grained solution architecture level and communicate better instead of more.

  - Make all elements easily realizable in solution design and implementation.

  - Allow some flexibility to meet a wide range of solution architectural needs.

- **Significant** architectural considerations:

  - Emphasize significant case scenarios, not a full solution architecture or complex model presentation.

  - Focus on key non-functional requirements (NFRs) and service-level characteristics and support decisional architecture with metrics and architectural walkthroughs.

  - Align with EA guides such as objectives, challenges, principles, and architecture building blocks.

---

[1] *Service* here means an abstracted component or concept, independent of implementation. It can also be referred to as an architectural service.

- **Systematic** liaison between different levels of architecture:

    – Adopt an IT service-based, enterprise-level solution architecture with a holistic approach to overlapping concerns in the architectural integration of EA, BA, AA, DA, and TA.[2]

    – Architect solution building blocks and their governing correlations, with less detailed descriptions but an effective reference to people, communication plans, project management, solution processes, and external environments.

Although many ideas exist in an enterprise solution, the A-ESA model plays a role as *a single source of truth* under the control of the ESA architect role. It represents the abstracted and influenced reality, created by the ESA architect and developed collaboratively by and for the stakeholders. The simple yet comprehensive modeling approach of A-ESA makes it easy and intuitive to use in practice.

Note that the popular term "Agile Architecture" refers to a general philosophy or framework in an Agile environment aimed at agility and time-to-market, while A-ESA applies to different architectural styles in a large solution environment leveraging an Agile approach. A-ESA covers Agile architecture in a customized way. It instills more architectural thinking for solution sustainability. See Appendix V for frequently asked questions and related comparisons.

The A-ESA Model Specification

A-ESA solution architecture is a collection of model elements and their relationships, coupled with architectural guidance and governance,

---

[2] They represent different levels of architecture: enterprise, business, application, data, and technology.

which represents a real solution system in an abstraction with different dimensions. The A-ESA model specification is an *ESA language* or a set of Enterprise Solution modeling notations that includes area, view, element, and property.

Architecture Area

An *area* is a grouping of relevant views. Areas can be of six types: (1) *enterprise capability area,* (2) *case scenario area,* (3) *outline overview area,* (4) *functional area,* (5) *operational area,* and (6) *validation area* (see Figure 1-1).

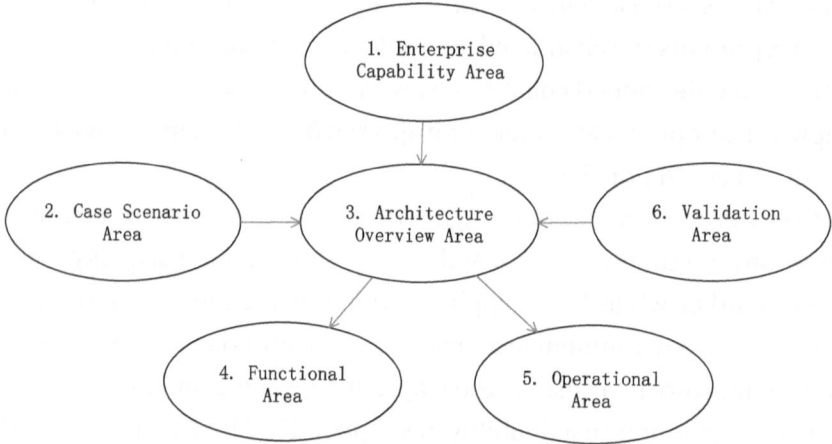

Figure 1-1. *Areas of the A-ESA model*

Area is a flexible definition. It can be represented slightly differently for each solution. For example, the functional area and the operational area are commonly referred to as the application area and the technology area, respectively. The validation area can be optional if cross-checks and walkthroughs are handled in other areas. Although the definition is flexible, areas must contain views with all the necessary elements to form a well-rounded architectural model.

Architecture View

The architecture view, the central piece of an Agile ESA model, addresses distinct stakeholder concerns about complex solutions within a *simplified view scope*. It is vital that each view is targeted to a specific audience or group of audiences from an intended viewpoint. No arcane technical information should be presented to the business user and vice versa for the technical specialist.

Table 1-1 shows the A-ESA fundamental views, and Table 1-2 is a list of view definitions.

Table 1-1. *Agile ESA Fundamental Views*

Area	Fundamental View	Primary Audience
Enterprise	Capability View	CxO, Strategy Consultant, Enterprise Architect
	Organization View	
Case Scenario	*Use Case Model/Process* or Page Flow View	User, Business Analyst
Architecture Overview	*Architecture Outline* (Landscape View, Roadmap, Migration View, Integration View, Cross-cutting View, etc.)	User, Lead Architect, ESA Architect
	Reference Architecture or Pattern	
	Metrics View	
	DevOps	

(*continued*)

Table 1-1. (*continued*)

Area	Fundamental View	Primary Audience
Functional	Service Interaction/Collaboration	Application Architect
	Service Relationship View	
	Information Relationship	
	Component Service Realization	
Operational	Package Mapping	Technical Solution Architect, Operational Manager
	Runtime Operational View	
Validation	Verification and Validation	Tester, Governance Body, ESA Architect

Note:

- *Italics* indicate a minimal or lean set of views.
- The Agile team can be any of these audiences.
- Architecture Outline View (architecture overview or introductory view) is the most flexible view that can take the form of an enterprise-level IT Environment View, Solution Context View (with project definition attributes: objective/scope/time/budget, etc.), sketch diagram (no correlation), and so on.
- Depending on the solution context, some views are optional, such as the Organization View and Service Component Realization View.
- Reference Architecture or Pattern View generally needs to be customized when expressed in a correlated solution view.
- The DevOps view cuts across functional and operational areas.
- Validation includes gap and impact validation, relationship cross-checking, and significant walkthrough validation. Validation can be thought of as part of the architecture overview. In fact, all architecture views are subject to validation.

Table 1-2. A List of View Definitions

Fundamental View	Prefix	Definition
Capability View	CAP	Provides a categorization and rationalization of capabilities.
Organization View	ORG	Represents the organizational impact.
Use Case Model View	UCM	Defines the interactions between the external actors and the system to achieve certain goals.
Process View	PRM	A collection of linked tasks that delivers a service or product to achieve a goal.
Page Flow View	PFV	Represents a storytelling GUI flow scenario.
Architecture Outline View	AOV	Takes distinctive forms and runs through different perspectives or aspects (application, technical, logical, physical, functional, or infrastructural). A high-level governing view containing a well-defined system context or solution building blocks.
Pattern View	PTN	Represents well-structured collaboration between IT services, and is a reusable asset.
Metrics View	MTS	The most vital view for architectural considerations, and its simple mapping supports decision-making.
DevOps View	DEV	Applies to Agile or evolutionary architecture for continuous development and deployment, and helps lay out the DevOps environment.
Service Interaction/ Collaboration View	SIV	Dynamic and typically reflects a specific case scenario.
Service Relationship View	SRV	Reflects the structure and modularity of services and their dependencies.
Information Relationship View	IRV	Reflects the structural relationship of information services.

(*continued*)

Table 1-2. (*continued*)

Fundamental View	Prefix	Definition
Component Service Realization	SCR	Describes how component services are realized in significant cases.
Package Mapping View	DPM	Maps each application level or functional service to a runtime package, taking into account its *unique architectural characteristics.*
Deployment/ Operational View	DEP	An operational view where middleware with deployment packages is associated with its node and network environment.
Validation View	VLD	Validates structural consistency, dynamic relationships, and decision behavior.

It is important to note that many of the A-ESA views focus on cross-cutting concerns, such as *outline view, metrics, DevOps, service relationship* (including UI, app logic, data, and tech elements), and *deployment package mapping.* This is in sharp contrast to the traditional layered solution architecture which has, for example, a clear separation between application architecture and technical architecture.

Note that the model in A-ESA is a composition of various views from different viewpoints. A habitually named use case model (UCM), for example, actually means one or more *views*, only part of the *holistic* A-ESA solution model.

Architecture Element

An *architecture element* is a set of iconography that describes and explores common concerns around the enterprise solution architecture. The role of a chief architect or lead architect in the field of enterprise IT system-level solution architecture is to address the architectural concerns and map them to the architecture elements.

Table 1-3 lists model elements that are classified as required IT service-based system elements, lean elements, base or basic elements, and standard elements. These element notations are essential in the context of solution architecture. Common element classifications are *lean*, *basic*, and *basic + assistive*.

- **IT software system elements** are the key elements for an enterprise system to function and operate.

- **A lean set of elements** constitutes a minimal solution architecture for simple yet common model presentation.

- **A basic set of elements** meets all the solution project needs.

- **Assistive elements** optionally provide a better visual representation and easier drill-down. Table I-3 lists some of the assistive elements.

- **A standard set of elements** includes base elements and commonly used assistive elements, as shown in Table 1-3.

Table 1-3. List of Elements

Category	Element	Prefix	Icon	System	Lean	Base
Enterprise	Capability	CP				X
	Value	VL				X
Case Scenario	Role	RO			X	X
	Task	TK				X
	Use Case	UC				X
Metrics	Principle	PR			X	X
	Requirement	RQ			X	X
	Key Choice	KC			X	X
	Risk	RK				X
	Governance	GV				X
Functional Service	UI Service	UI		X		X
	App Logic Service	AS		X		X
	Data Service	DS		X		X
	Tech Service	TS		X		X
	Service Interface	SI		X		X
	Service Component	SC		X	X	X

(continued)

Table 1-3. (*continued*)

Category	Element	Prefix	Icon	System	Lean	Base
Operational Service	Deployment Package Unit	DP		X		X
	Middleware	MW		X	X	X
	Database	DB			X	X
	System Device	SY		X	X	X
	Node	ND		X	X	X
	Network	NW		X		X
	Location	LO		X		X
Connection	Association	AN	——	X	X	X
	Flow	FW	⟶		X	X
	Composition	CN	—●		·	X
	Realization	RN	----▷			X

(*continued*)

Table 1-3. (*continued*)

Category	Element	Prefix	Icon	System	Lean	Base
General	Note	NT				X
	Grouping	GP			X	X
	Generic Domain	DM			X	X
	Generic Service	GS			X	X
	Virtual Service	VS				X
	View Frame	VF				X
	Deliverable	DL				X
	Artifact	AR			X	X
	Extension	EX				X

(*continued*)

Table 1-3. (*continued*)

Category	Element	Prefix	Icon	System	Lean	Base
Assistive	Frontend	FE				
	Application	AP				
	Process/Composite	PS				
	Microservice	MS				
	Cloud	CL				
	Queue	QU				
	Event Service	ES				
	Item	II				

Note:

— A-ESA is more outcome-driven, so the high-level, intent-oriented elements (goal, outcome, resource, etc.) are used sparingly. These intent-oriented elements will be part of the IT strategic architecture and they must be specific and relevant when referenced in the enterprise solution architecture.

— Five architectural *metrics* or criteria are intended to ultimately satisfy the requirements metric. They also serve as metrics for each other or for other elements.

— The database is generally considered middleware, but it's so dominant in solution architecture views that it's included as part of the foundational or basic elements.

— Some elements are from a solution management perspective, such as value/cost and deliverable.

— This table is adapted from *Agile Enterprise Solution Architecture*, by Gu, Sean, Vernal Press, 2021

A-ESA adopts a minimal but representative set of elements that practically fit the enterprise solution architecture in terms of their unique characteristics. In the IT world, different terms can mean the same thing, and the same concept can be expressed in different terms. Therefore, clearly specifying the nature of an element (using the *element prefix*) is extremely important in solution definitions.

For a general solution architecture, basic or standard elements should suffice. For more elaborate A-ESA modeling, you can use derivative elements (such as non-functional requirements [RQ-NFR]), additional assistive elements or special images (such as AWS cloud notations).

Note that derivative elements and special images are not additional element types, but naming conventions or display notations (see Table I-7 in Appendix I for naming conventions). This partially embodies the Agility feature of A-ESA modeling. However, for the sake of simplicity and consistency, each element, whatever it's called, must correspond to one of the basic elements of A-ESA.

The A-ESA model specifications are not covered in detail in this book. However, refer to Appendix I for a collective list of elements applicable to various architectural styles. All of the model elements described in this book are listed there. If you need more than what is listed, your architecture model is probably too complex and beyond the scope of the enterprise solution architecture.

Architecture Property

Property is a set of *name-description* pair[3] *attributes* or *parameters* associated with a model, view, or element, and it can contain sub-properties if needed. It is a hardcore part of the architecture model and records important specifications, measurements, and special

---

[3] It can also be any other pair, such as property-description, name-value, attribute-value, name-type-value, etc.

considerations. It makes architectural modeling richer, yet simpler and cleaner. A mere architecture model without key property specifications, especially for significant elements, is generally hard to measure.

A-ESA Lean Model Mode

For small or fast solutions, you can use a lean set of areas or views and a lean set of elements (see the indicators in Tables 1-1 and 1-3, respectively) in a minimally simplified approach. The number of lean elements is reduced to almost a third of the base elements. This enables a quick learning curve but still maintains a relatively full view of the solution architecture.

The lean areas/views and elements can be mapped to the basic model specifications, as shown in Tables 1-4 and 1-5. The coarse-grained notations can achieve a similar A-ESA result. For example, the *generic service* (GS) element can cover various functional services, using the *element prefix* or *property attribute* to indicate the service type if necessary; the *system* (SY) element can represent such assistive elements as *mobile device or frontend PC,* and so on.

Table 1-4. *Lean View Notations and Their Coverage*

Lean View Area	Coverage View
Case Scenario	*Use Case Model/Process View*
	Page Flow View
Outline	Capability View
	Architecture Overview (Landscape, Solution Context, IT Environment, Migration View, Integration View, Cross-cutting View, etc.)
	Reference Architecture or Pattern View
	Metrics View
	DevOps View
	Overall Validation View
Application/Functional	*Service Collaboration View*
	Service Relationship View
	Information Relationship View
Technical/Operational	Package Mapping View
	Operational Deployment View
Note:	
— The italics indicate the lean views in each view area.	

Table 1-5. *Lean Element Notations and Their Coverage*

Category	Lean Element	Additional Coverage
Metrics	Key Choice	Risk
		Governance
	Principle	Value
	Requirement	Task
		Use Case
		Value
Functional	Component	Service Component
		Service Interface
	Database	Data Service
	Generic Service	UI Service
		Capability
		App Logic Service
		Data Service
		Tech Service
		Service Interface
		Deployment Package Unit
		Process
		Microservice

(continued)

Table 1-5. (*continued*)

Category	Lean Element	Additional Coverage
Operational	Middleware	Network
	Node	Network
	System	System/Device
General	Role	User/Actor
	Artifact	Deliverable
		Extension
	Generic Domain	Location
		Composition
		Grouping
		Capability
		View Frame

Note:
- As can be seen from the table, the lean element mode has the least impact on the operational elements.
- A lean element can mean different coverage depending on the solution environment.
- A slight adjustment is likely, for example, to replace the *component* with the *service interface* for an API-driven architecture, or include the *task* element to better represent the case scenario.

Figure 1-2 shows a visual list of the primary lean A-ESA element notations. It's an attractive option for IT folks to quickly see the solution architecture in action. It can be used for initial enterprise solution modeling. In most cases, the list is sufficient for a quick and meaningful A-ESA model. IT architects, especially Agile users, are encouraged to use the lean A-ESA mode.

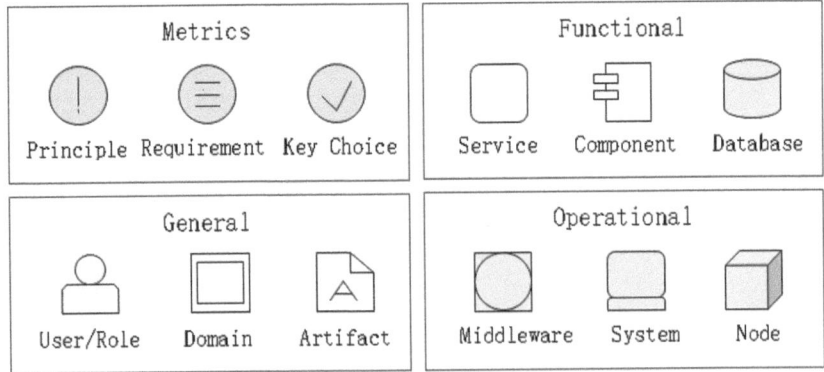

Figure 1-2. *Lean mode A-ESA elements*

Note that the lean model mode is not sufficient for the more purposeful ESA, such as architectural design guidance, deployment mapping, and division of labor. Chapter 2 primarily uses a base set of elements to better reflect ESA model deliverables in various solution profiles. Of course, the base or assistive elements can be easily converted to lean mode elements using the naming convention or tool support.

Summary

This chapter briefly introduced the intent and modeling specifications of A-ESA. It is the foundation of various enterprise solution architectures, and it is the result of careful consideration of proven solutions and the six constituents of A-ESA. The following chapters discuss how to apply and map the specification to the solution environment. The next chapter starts with eight selected solution cases using A-ESA notations.

CHAPTER 2

A-ESA Modeling by Example

As an ESA architect, think always, model often, and implement[1] sometimes.

A-ESA[2], briefly introduced in Chapter 1, follows the principle of Occam's razor, which favors simple solutions because modeling complexity is more error-prone.

It is often said that a picture is worth a thousand words, and in this chapter, the example solutions are presented with fewer words. Proper visualization as well as concise property specifications are the keys to architectural modeling. While modeling a solution, keep in mind the following:

- Don't go into detail, because abstraction is one of the key techniques in IT architecture. Enterprise solution architecture creates IT service-based building blocks, a bit of a black-box effect.

---

[1] *Implement* means learning new coding skills and middleware technologies, making prototypes (either proof of concept or proof of technology), providing implementation samples and guidance, etc.

[2] A-ESA is also a short name for an A-ESA Model.

© Sean (Chunhong) Gu 2024
S. Gu, *Mastering Enterprise Solution Modeling*,
https://doi.org/10.1007/979-8-8688-0992-7_2

- Focus on significant scenarios, processes, value streams, or throughputs that may require service component realization views or walkthrough views for solution assurance.

- Don't blindly replicate someone else's architecture. Think through and iteratively build an architecture that better fits the non-functional requirements and its changing needs.

This chapter serves two purposes: to familiarize you with Agile ESA modeling, and to provide you with an initial understanding of architectural thinking and measurement using common architectural styles and techniques (which will be discussed further in subsequent chapters). The A-ESA examples in this chapter include eight enterprise solution architecture (ESA) profiles (see Table 2-1), based on an initial *order processing system* (OPS) solution architecture,[3] which is part of the order management system (OMS). The OMS case scenarios are relatively straightforward and easy to comprehend.

---

[3] The initial solution is detailed in my book *Agile Enterprise Solution Architecture*, which includes the project background, issues/needs, and architecture/governance views based on a BRS retail case.

Table 2-1. *Example Solution Profiles*

#	A-ESA Profile	Architectural Focus	Major Role
1	Strategic Architecture	ESA principles and capabilities	– Chief Architect
2	Business Architecture	IT-based business capability and scenario mapping	– Business Architect
3	Reference Architecture	Architectural pattern views from vetted solutions and industry practices	– Lead Architect
4	Enterprise Integration Architecture	Primarily as an enterprise application integration architecture that can be used as a reference architecture	– Lead Architect – Integration Architect
5	Application Governance Architecture	Application architecture considering functional techniques and governance	– Application Architect
6	Microservice Architecture	Covering bounded context identification and operational environment	– DevOps Architect
7	Cloud Hosted Architecture	As an add-on application hosted on the cloud platform	– Cloud Solution Architect
8	Big Data Architecture	As a cross-cutting technical architecture with big data considerations	– Data Architect – Technical Architect

The example solutions do not focus on *case sharing*, although they are based on leading practices and real cases. The examples focus on element selection and placement and demonstrate some notation usages and property specifications to facilitate architectural thinking. They are intentionally simplified, just enough to capture the salient aspects of key architectural considerations without much domain knowledge, business expertise, or specific technical competence. In other words, for a specific full-blown solution project, the organization and presentation of the entire model would have been more relevant and enhanced on top of these profiles. Hopefully, this streamlined yet practical approach will help you gain a good command of A-ESA architectural skills.

Note that for the assistive element images in the profiles that are not specified or listed in Table 1-3, refer to Table I-5 in Appendix I. The customized notations help present a familiar look and feel for better elaboration and human understanding. Nevertheless, all custom element notations are bounded within the A-ESA base or basic element category.

HINT[4]

Multiple Profiles: The eight profiles are relatively independent for clearer modeling demonstration. Enterprise solution architecture often includes more than one architectural style and different architectural layers. A chief/lead ESA architect takes charge of the overall solution modeling with multiple profiles.

---

[4] Hint: Answering common questions from junior architects during the A-ESA modeling process.

A-ESA Profile 1: Strategic Architecture

For A-ESA, strategic architecture will be driven down to architecture principles, values and capabilities, reflecting the intent of all IT strategic plans. The strategic architecture will require effective communication with enterprise business leaders and will provide abstract guidance to enterprise solutions. The strategic architecture usually *won't be a standalone architecture*, and in fact it can be part of the A-ESA metrics view mapping process or included in another solution profile.

Note that this profile represents an initial strategic IT architecture that includes only principles and capabilities and requires further elaboration.

ESA Principles

A *principle* can convey tons of words and reflects an enterprise's strategic intent from multiple sources. For the *scalable online order management system*, you must first identify the *guiding principles* at the enterprise solution level, followed by the *solution principles* in the four basic architectures: business architecture, application architecture, data architecture, and technical architecture.

Guiding Principles

A *guiding principle* is one that dominates the entire enterprise solution. Only a select few will be defined. Figures 2-1 and 2-2 show the *guiding principles* that direct the *solution principles* in a high availability solution.

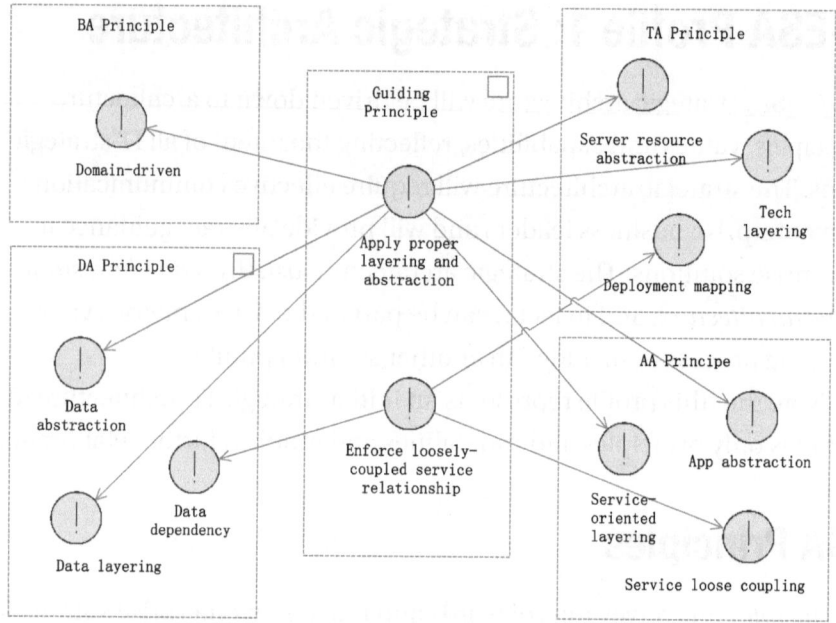

Figure 2-1. *Guiding principles (1)*

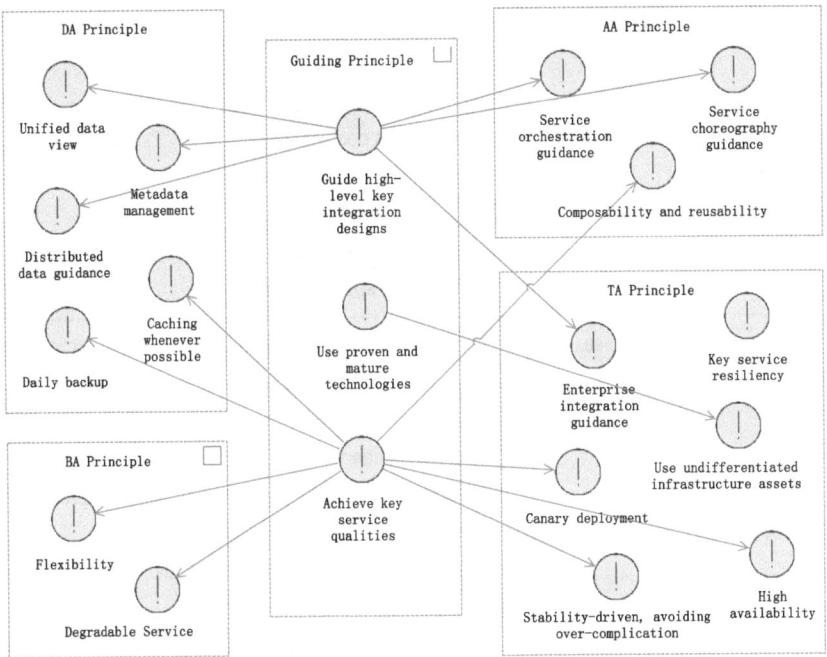

Figure 2-2. *Guiding principles (2)*

HINT

Subgrouping: If there are many elements in a view that appears cluttered, you can subdivide the elements using several approaches, such as multiple views (as seen in Figures 2-1 and 2-2), view frame element, element layering or grouping, tree view, or tabular view. Remember that the purpose of any view element is for correlation and visual validation. Insignificant elements can be removed from the view or recorded in the view properties or attached documents.

Solution Principles

Table 2-2 shows the *solution principles* for each of the four basic architectures mapped in Figures 2-1 and 2-2. These solution principles can be reformulated to better reflect their focus, and the solution principle description elements can be further mapped to non-functional requirement (NFR) metrics for key considerations.

Table 2-2. *Properties of Solution Principles*

Principle	Property	Description
BA Principle	Domain-driven	– Core business domain split-up – Business process composition – Service-level categorization – Business platformization – Business rule specification
AA Principle	Service loose coupling	– Functional services not dependent on component realization – Functional services not directly dependent on underlying servers – Avoid codependency – Functional services not dependent on business process services – Whenever possible, async communication between app domains – Async communication for non-core services and between core and non-core services – Mid-platform services not dependent on upper-level functional services
	Service-oriented layering	– Stable vs. instable app services – Core vs. non-core app services – Main vs. sub business processes – App services vs. data services – App services vs. technical services – App services vs. UI services – Functional servicers vs. services components
	App abstraction	– Service domain abstraction – App clustering abstraction
	Service encapsulation	– API specification – Stateless whenever possible

(continued)

Table 2-2. (*continued*)

Principle	Property	Description
DA Principle	Data abstraction	– App system only dependent on logical DB for data store abstraction – App system access to DB via service interface for data access abstraction
	Data layering	– Read/write separation by data access frequency – Data access volume – Business service-level – Important or special business data – Dynamic/static layering by hot/cold data, history data – Tech and utility vs. app and data services – Data sharding for HA by unique ID – Database or schema design by business domain
	Data dependency	– Database not dependent on location and sharding
TA Principle	Tech layering	– Zone by business – Read for replication vs. write for HA
	Deployment mapping	– NFR technical or utility services not dependent on app or data services – Deployment package matching overall SLA considerations
	Key service resiliency	– Timeout or queue limit for sync calls – Clustering for fault-tolerance – Runtime self-governance
	High availability	– N+1 design plus DR – Active/active, multiple data center deployment

Capability Views

Figure 2-3 is a subset of the *retail business capability model,* an integral part of business architecture.

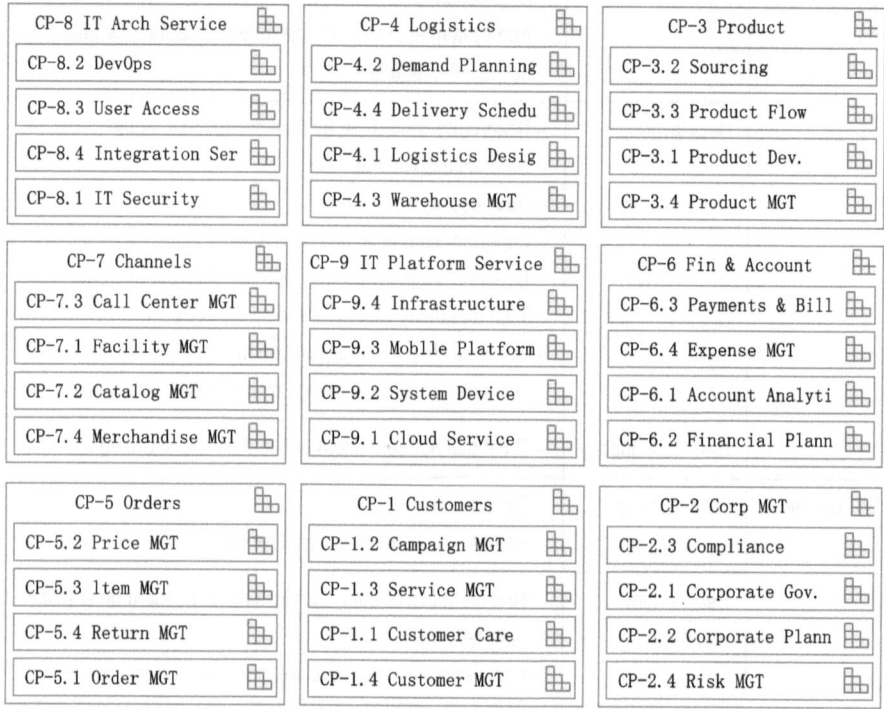

Figure 2-3. *Retail capability model*

Figure 2-4 shows a solution level capability view focusing on the distributed order management system (OMS). As you can see, the solution capabilities are roughly expressed using various functional and operational services that are to be built on the cloud service platform.

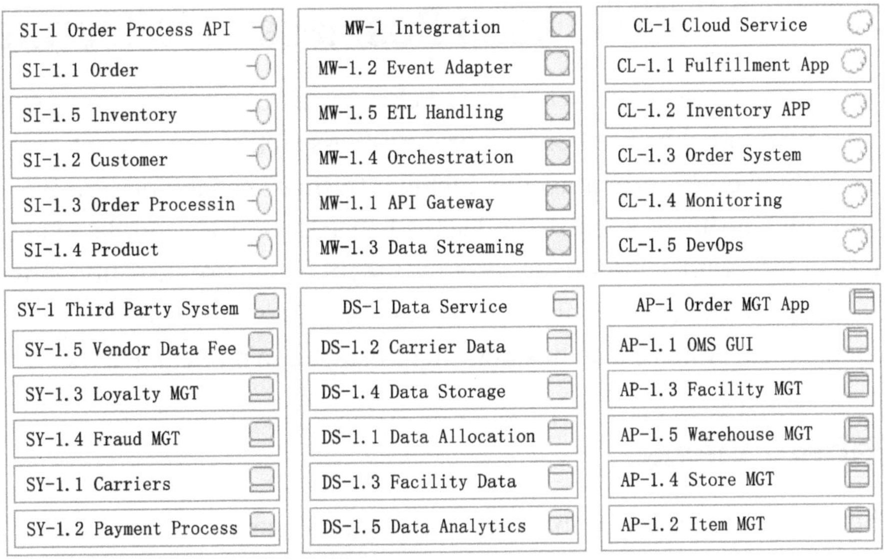

SI-1 Order Process API	MW-1 Integration	CL-1 Cloud Service
SI-1.1 Order	MW-1.2 Event Adapter	CL-1.1 Fulfillment App
SI-1.5 Inventory	MW-1.5 ETL Handling	CL-1.2 Inventory APP
SI-1.2 Customer	MW-1.4 Orchestration	CL-1.3 Order System
SI-1.3 Order Processin	MW-1.1 API Gateway	CL-1.4 Monitoring
SI-1.4 Product	MW-1.3 Data Streaming	CL-1.5 DevOps

SY-1 Third Party System	DS-1 Data Service	AP-1 Order MGT App
SY-1.5 Vendor Data Fee	DS-1.2 Carrier Data	AP-1.1 OMS GUI
SY-1.3 Loyalty MGT	DS-1.4 Data Storage	AP-1.3 Facility MGT
SY-1.4 Fraud MGT	DS-1.1 Data Allocation	AP-1.5 Warehouse MGT
SY-1.1 Carriers	DS-1.3 Facility Data	AP-1.4 Store MGT
SY-1.2 Payment Process	DS-1.5 Data Analytics	AP-1.2 Item MGT

Figure 2-4. Distributed OMS solution capability view

A-ESA Profile 2: Lean Business Architecture

This profile demonstrates a business architecture from an ESA perspective and is therefore different from a "pure" business architecture model. This streamlined business architecture includes business capabilities (and associated business processes and activities) and is closely correlated to IT architecture, with value stream resource mapping and use case modeling.

Note that this profile contains only a small part of an OMS (order management system) lean business architecture as a modeling demo.

Pattern Views

The business architecture in an enterprise solution architecture context includes strategic and operational interface content, so its coverage differs from traditional business analysis.

Figure 2-5 is a *metadata pattern* for the lean business architecture (or business information architecture), where close considerations are given to IT and business. The architecture is divided into four sub-models with simplified representations and minimal notations. Note that the notational representations and model divisions may be different to suit the needs of each solution such as automation, digitization, big data solution, and so on.

HINT

A-ESA Concerns in Business Architecture: In a traditional modeling approach, the business information architecture is complex, but the Agile ESA forces architects to think abstractly and focus on the architectural concerns. For example, for capability model, what are the key principles that map to business and data architecture? How does the heatmap analysis and value stream impact the service level requirements? In addition, A-ESA is for solution mapping, not business analysis or data modeling. So do not delve too deeply into such familiar terms as atomic subject area, atomic data element, business item, enumeration item, E/R diagram, or even type or attribute, although they can be recorded as property specifications if necessary.

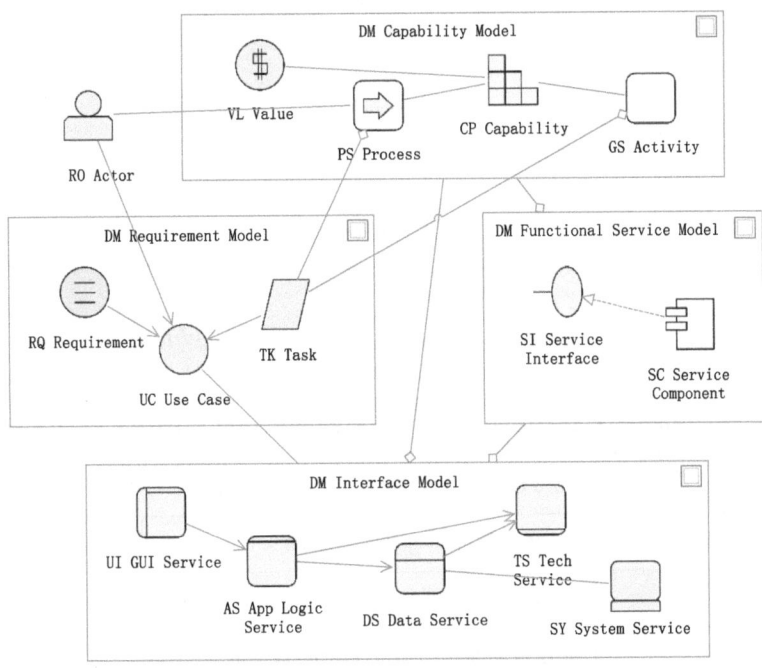

Figure 2-5. *IT-based lean business architecture*

Use Case Model Views

Figure 2-6 shows a traditional customer management use case model (part of the order management process) *view* that expresses how the user and system interact. Table 2-3 shows its *search customer* use case property, which seems old-fashioned, but is still used effectively for reliable large-scale system solutions.

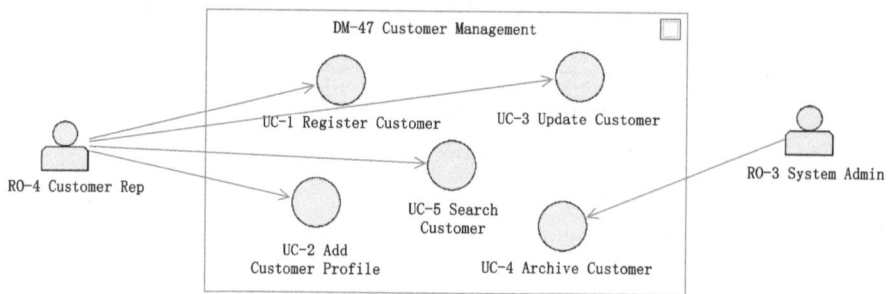

Figure 2-6. *UCM-2 customer management*

Table 2-3. *Use Case Property of Search Customer*

Property	UC-5 Search Customer
Scope	Online order management system
Objective	Search registered customer information
Precondition	Actor is authorized to access the system
Success Outcome	Searched customer information is obtained
Failure Outcome	Searched customer information is not found
	Searched customer information is not detailed
Primary Actor	Customer Rep
Secondary Actor	System Admin
Main Scenario	1. System prompts actor for search criteria: either customer name or social security number
	2. Actor provides the required info
	3. The searched customer info is shown
	4. Use case is successful

(*continued*)

Table 2-3. (*continued*)

Property	UC-5 Search Customer
Alternatives	2.1 The search criteria are not found
	2.1.1 System displays message: no matching info is found
	2.1.2 Use case fails
	2.2 Multiple customer records are found
	2.2.1 System shows a list of matching records
	2.2.2 Actor chooses from the list
	2.2.3 Use case continues to Step 3
Variations	None

Business Process Views

Figure 2-7 is one of the process views for the order management solution. Note that only a few business processes that have a significant impact on business are considered for A-ESA. Table 2-4 shows property attributes that capture key business concerns of the *order confirmation process view*.

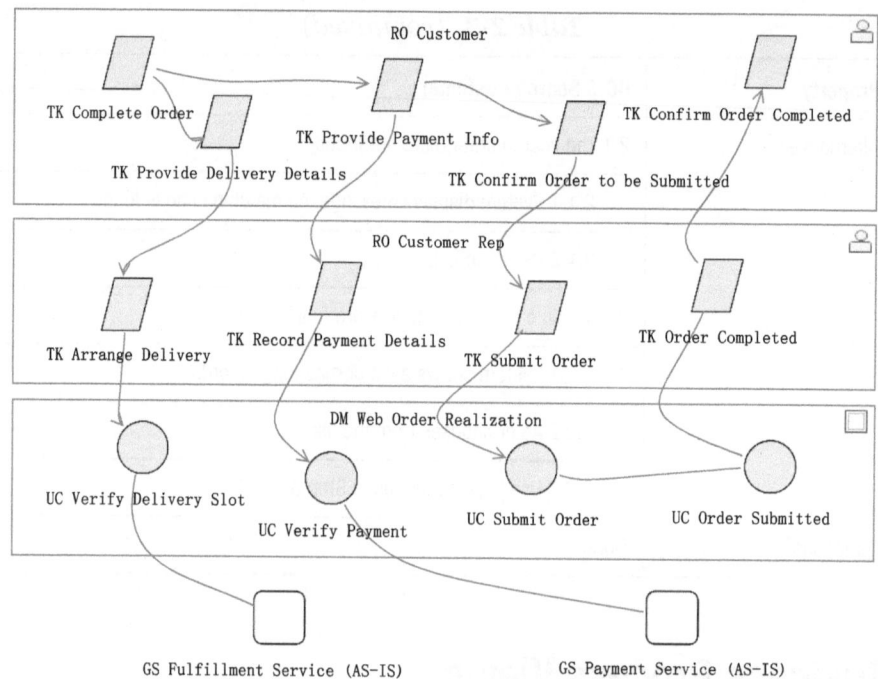

Figure 2-7. *Order confirmation process view*

Table 2-4. *Properties of the Order Confirmation Process*

View	Property	Description
PRM-4 Order Confirmation Process View	Name	Order Confirmation
	Type	Business Process
	Role	Customer
	Role	Customer Representative
	InputData	Shopping Cart Data
	OutputData	Submitted Order Data
	KPI	Processing Time
	ITRequirement	– Optimize order processing workflows for improved efficiency and reduced time – Avoid single point of failure from order initiation to order confirmation – Enable automated notifications and real-time tracking – Avoid manual data entry errors with meaningful warning messages
	RequirementSource	Business Group I AS-IS Architecture

HINT

Process View vs. BPMN: ESA uses only roles, tasks, and connecting lines, a streamlined process approach, for architectural thinking. BPMN (Business Process Model and Notation) involves a more detailed and formal approach and is too complex for ESA. For justified significant cases, you can work with design/development team using BPMN, even execution workflow or process language such as BPEL (Business Process Execution Language) for an end-to-end mapping.

Process View Mapping with Value Stream

Figure 2-8 is an illustrative *order processing application* value stream with four stages. The value stream reflects its stakeholder, trigger point, value created, and value items (see Table 2-5).

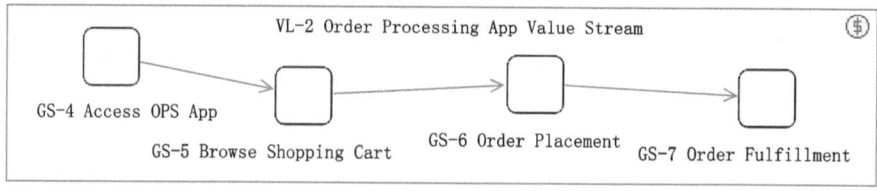

Figure 2-8. *Order processing app value stream*

Table 2-5. *Element Property of the Value Stream*

A-ESA Notation	Property	Description
VL-2 Order Processing App Value Stream	Stakeholder	An online shopper
	Trigger	The customer places an order online
	Value Creation	The customer has a good shopping experience in terms of usability and availability
GS-4 Access OPS (Order Processing Solution) App	Entry Criteria	If the customer is already logged in, skip the identity check
	Exit Criteria	The customer is logged in to OPS App
	Value Item	Secured login

Figure 2-9 shows the mapping between value stream and generic service[5] capabilities. The capabilities within a value stream stage are defined as service capability instances. Modeling the capability instances will enhance the capability architecture model through cross-referencing and optimization.

---

[5] Generic services can be treated as business services by non-technical people.

Figure 2-9 also shows further mapping of the value stream capability instance to resources and functional/process services.

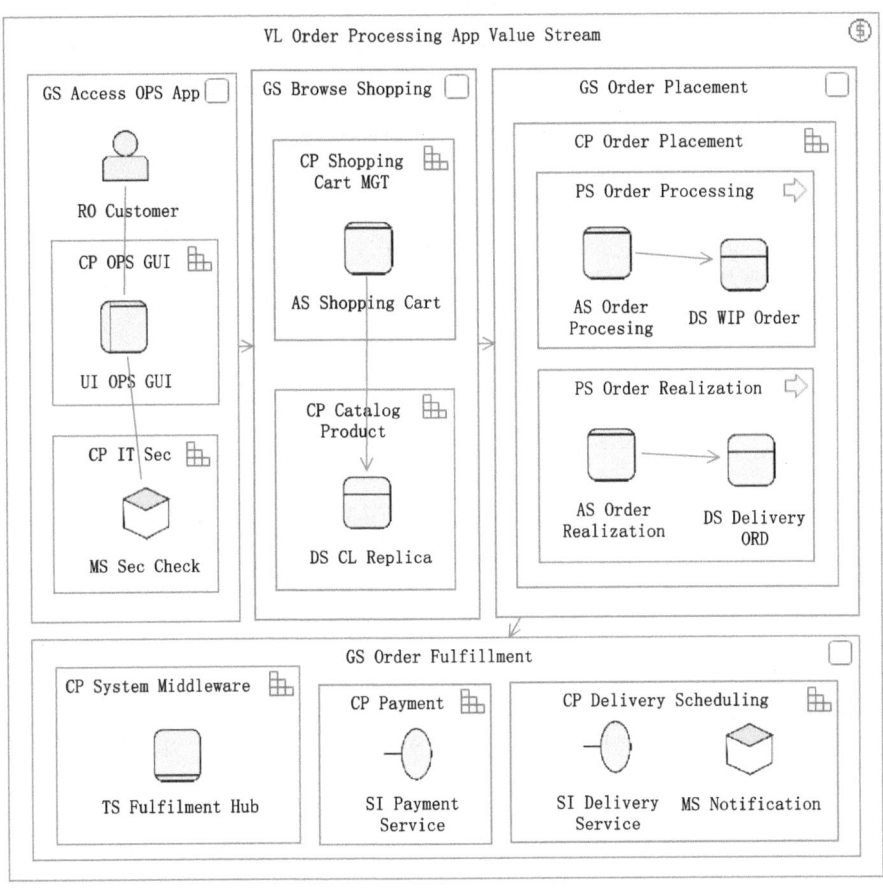

Figure 2-9. *Value stream mapping with capabilities*

HINT

Verification of Capabilities: One of the objectives of this detailed level of value stream mapping is to verify the capability model. Through an iterative process, capabilities are added, improved, optimized, or retired.

Architecture Outline View: Service Layering

This section illustrates a common service mid-platform intended for *packaged business capability platform domain* with shared service offerings. Figure 2-10 shows three layers of service from a business perspective: frontend interactive service, mid-platform service, and backend support service.

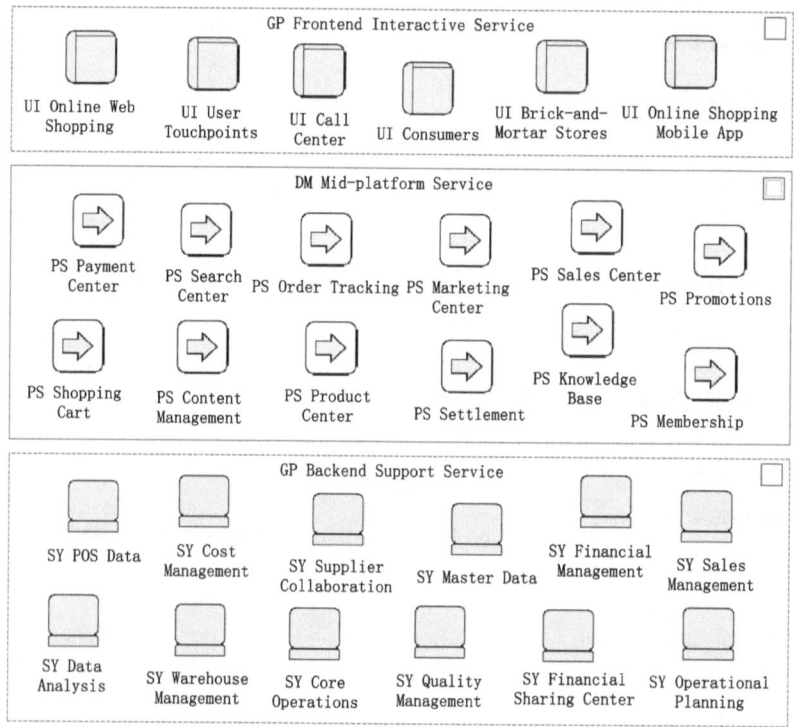

Figure 2-10. *Business capability platform*

HINT

Packaged Business Capability:[6] You can define services from different perspectives, such as reusability by user, catalog, promotion, product, search, and payment.

---

[6] Commonly referred to as mid-platform services

High-level Functional Service Views

Figure 2-11 shows a business architecture view in the form of functional services including UI services, application logic services, data services, technical services, and processing services. Since the missing link between business architecture and application architecture is another main architectural issue, the functional service outline view plays a pivotal role in clarifying business intent and business collaboration.

TS-1 Technical Service	AS-1 Application Service	DS-1 Informaton Service
TS-1.10 Private Cloud	AS-1.1 Order Service	DS-1.4 Account
TS-1.3 Data Center	AS-1.6 Transit Servic	DS-1.5 Promotion
TS-1.9 Monitoring Plat	AS-1.9 Quota Service	DS-1.8 Catalog
TS-1.4 Unified API Gat	AS-1.10 API Access Ser	DS-1.7 Product
TS-1.8 Open Platform	AS-1.8 Ledger Service	DS-1.9 Order
TS-1.6 Payment Gateway	AS-1.7 Pickup Service	DS-1.1 Customer
TS-1.1 Web Search	AS-1.4 Delivery Slot S	DS-1.2 Registration
TS-1.7 Logging Platfor	AS-1.3 Customer Relati	DS-1.6 VIP Service
TS-1.2 Image System	AS-1.5 Sorting Service	DS-1.3 Login
TS-1.5 Mobile Gateway	AS-1.2 Order Taking Se	DS-1.10 Pricing

PS-1 Business Process	RL-1 Buiness Rule Serv	UI-1 Interactive Service
PS-1.3 Order Input Pro	RL-1.10 Payable Receiv	UI-1.7 Order Service P
PS-1.5 Order Refund pr	RL-1.9 Search	UI-1.8 Payment Page
PS-1.10 Order Splittin	RL-1.8 Transit	UI-1.2 Single Web Page
PS-1.7 Shopping Cart S	RL-1.7 Refund	UI-1.6 Shopping Cart P
PS-1.6 Order Cancellat	RL-1.1 Order Splitting	UI-1.5 Order Menu Page
PS-1.4 Order fulfillme	RL-1.6 OrderLabel	UI-1.3 Listing Page
PS-1.2 SecKill Process	RL-1.2 Order Transfer	UI-1.4 Channel Page
PS-1.1 Order Process S	RL-1.5 SecKill Sales	UI-1.1 Front Page
PS-1.8 Payment Service	RL-1.3 Risk Control	UI-1.9 Confirmation Pa
PS-1.9 Transfer Servic	RL-1.4 Presale	UI-1.10 Search Page

Figure 2-11. *Business architecture: high-level functional services*

Functional Service Interface Views

Figure 2-12 shows a business service interface view. Careful identification and structuring of business services will be critical to the application of APIs later.

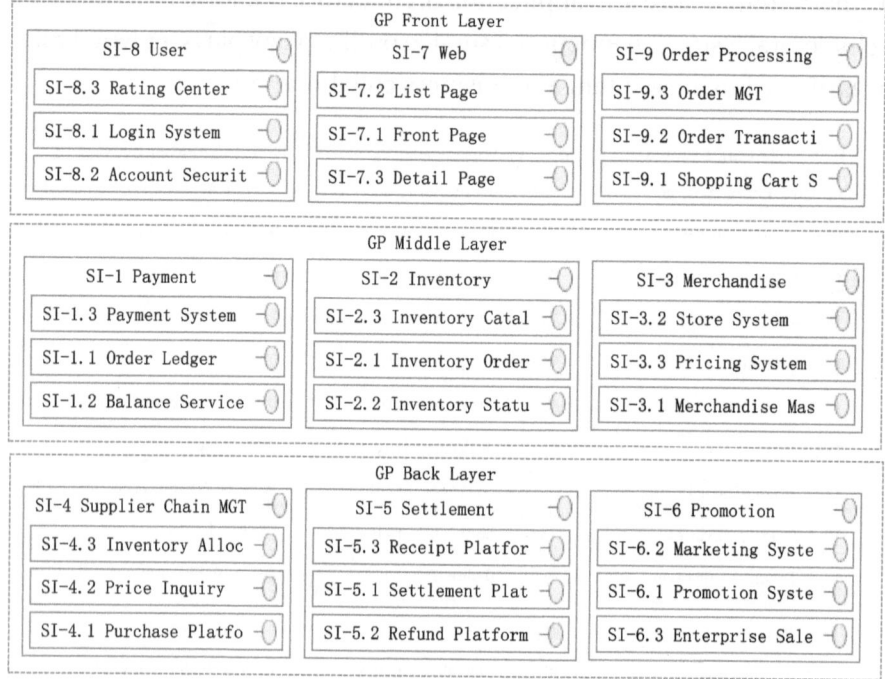

Figure 2-12. *Business architecture: service interface view*

A-ESA Profile 3: Reference Architecture

A-ESA pattern views refer to both the reference architecture (RA) and the pattern. While a pattern typically focuses on a single recurring solution problem or a subset of solution, a reference architecture usually reflects the target solution as a whole. This profile describes some commonly used reference architectures that can be applied to a related solution profile.

Note that further elaboration beyond this one-page view is required for better reference architecture guidance.

HINT

RA vs. Asset: A solution reference architecture is typically a set of views and documents that guide ESA modeling. The high-level RA views illustrated in this section can be elaborated into a set of solution views that fit your environment. Do not reinvent the wheel when a well-established reference architecture can be leveraged and customized. On the other hand, your well-tested solution can also become a reference architecture (as seen in Profile 4) in a defined environment. At the commercial level, many ISVs offer reference architectures that include components with solution services and solution codes, solution resources, and solution interaction interfaces and integrations. Some ISVs own their complete set of iconic notations for RA representations.

Application Business-centric Reference Architecture

Figure 2-13 is a reference architecture view for the digital order management solution.

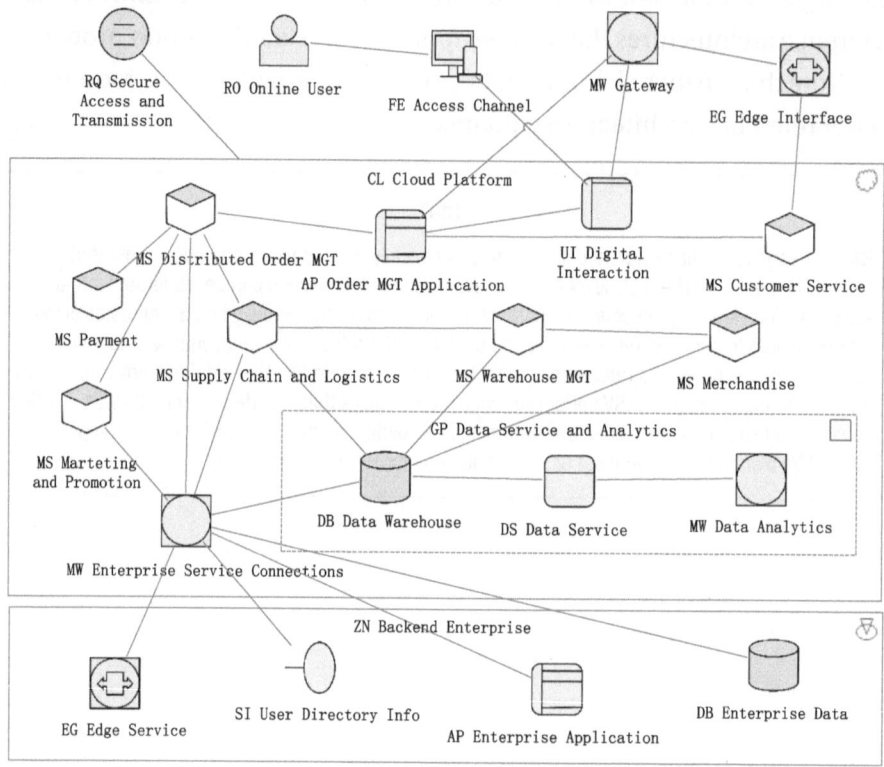

Figure 2-13. *Reference architecture of digital order management*

Data-centric Reference Architecture

Figure 2-14 is an architecture reference/pattern view for data analytics.

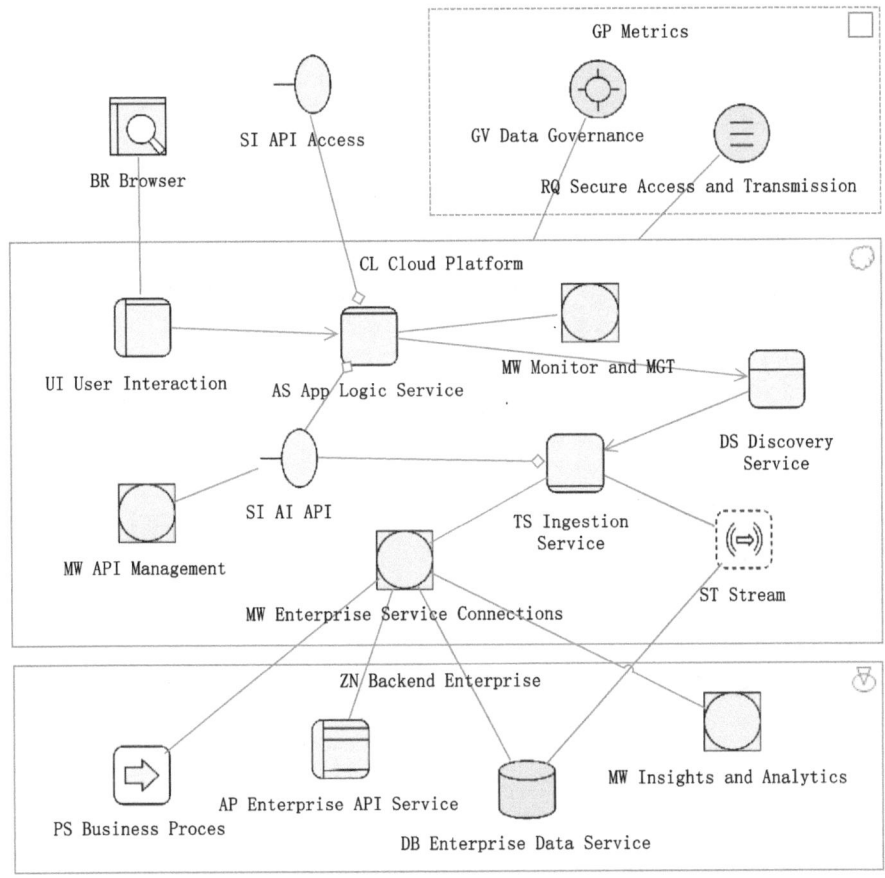

Figure 2-14. *Data analytics reference architecture*

Technology-centric Reference Architecture

Figure 2-15 shows an infrastructure reference architecture view.

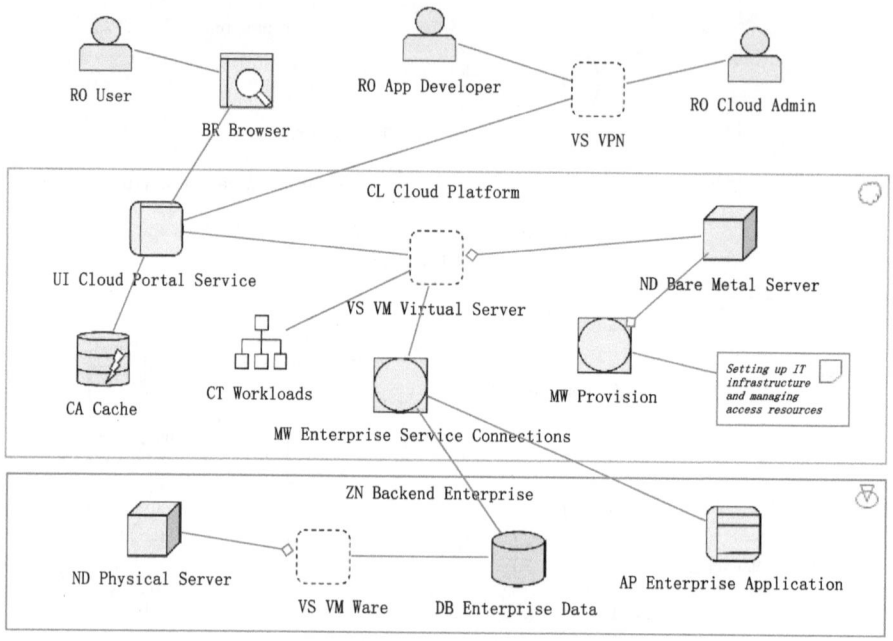

Figure 2-15. *Infrastructure reference architecture view*

Context-based Reference Architecture

Context-based reference architecture is a recommended structure for complex systems that must operate in a variety of environments or conditions.

Figure 2-16 shows an order management process system (OPS) architecture view adapted from a *cell-based reference architecture*, where a cell represents a coarser-grained domain-oriented services or a microservice of microservices. See Table 5-26 in Chapter 5 for more information on this architectural style.

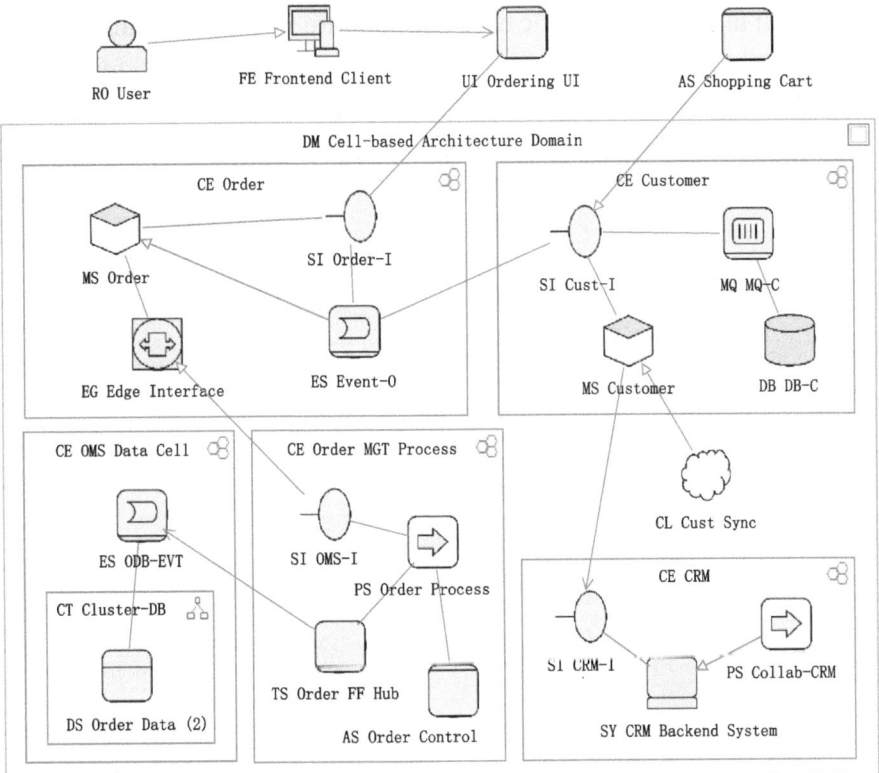

Figure 2-16. *OPS in a cell-based reference architecture*

Baseline Reference Architecture

The baseline reference architecture is generic, holistic, or high-level, rather than targeting a specific base architecture (BA, AA, DA, and TA). Figure 2-17 highlights a nine-block *cloud platform reference architecture* for a large retailer, with each block having an ESA focus. Figure 2-18 profiles a partial view of a baseline *chat AI reference architecture*[7] with the relevant ESA notations.

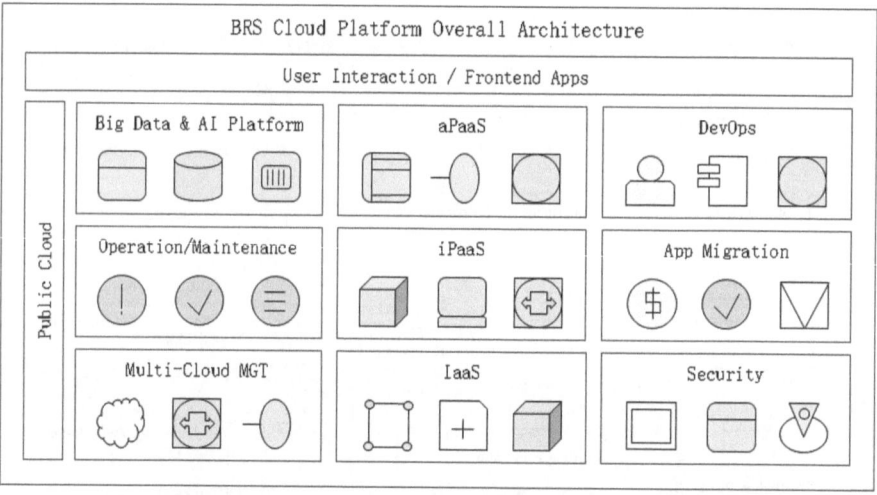

Figure 2-17. *Enterprise cloud platform reference architecture*

---

[7] Adapted from Microsoft's baseline OpenAI end-to-end chat reference architecture.

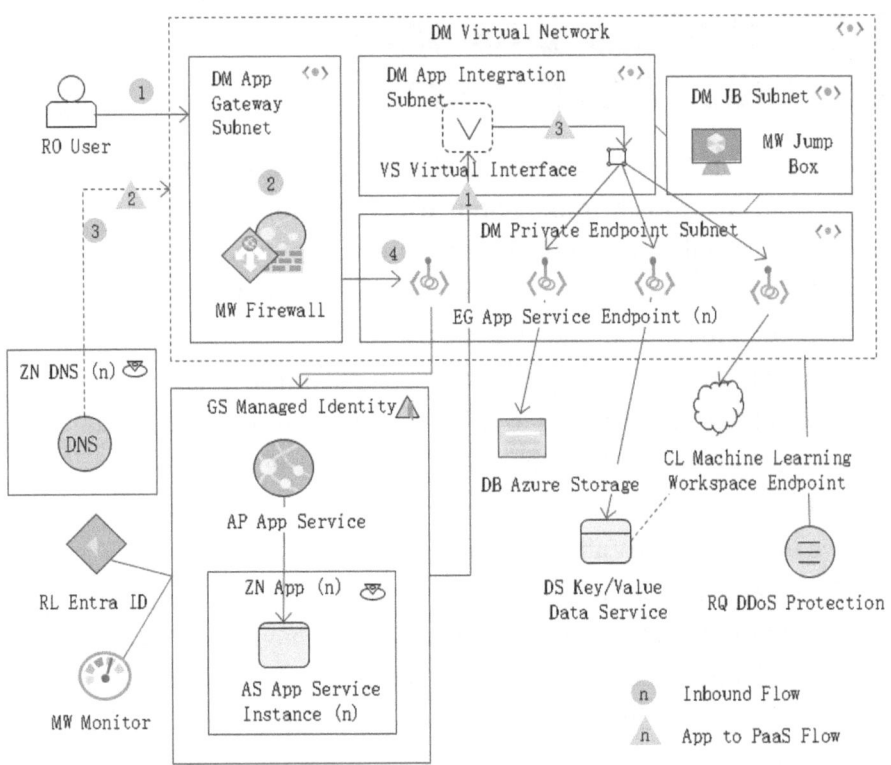

Figure 2-18. *Baseline chat AI reference architecture*

A-ESA Profile 4: Enterprise Integration Architecture

The *enterprise integration service* (EIS) solution is built around a service bus architecture, mostly from reference architecture considerations. It is virtually an Agile integration architecture, partially built on the cloud-native integration infrastructure, allowing fine-grained integration deployment and collaboration via APIs. It is a hard solution because it is intended to be a platform product for general purpose connectivity. The focus of this illustration is to clarify capability model, technical requirement mapping, domain service drilldown views, properties of the integration service interface applicable to different service contracts, to achieve *interoperability* and *reusability*.

Note that the intended audience for this profile is the integration architect in the specific solution context. Others may need more background information for better understanding. With the business audience in mind, a different set of views could have been presented for this solution.

This solution architecture used some special display images of the *general service* and *middleware* elements (see Table 2-6).

Table 2-6. _Custom Element Images for the Solution_

Prefix	Element Type	Element Name	Image
MW	Middleware	Message Queue Manager	
MW	Middleware	Message Flow Engine	
MW	Middleware	Wrapper Registration Manager	
MW	Middleware	Core Gate	
MW	Middleware	Component Manager	
GS	Generic Service	Industrial Identification Parser	
GS	Generic Service	Service Bus	
GS	Generic Service	Admin Shell	

Capability Views

The following are the Level 1 (Figure 2-19) and Level 3 (Figure 2-20) views of the EIS capability model. These capabilities are derived from enterprise-level IT strategic planning and proprietary project implementations.

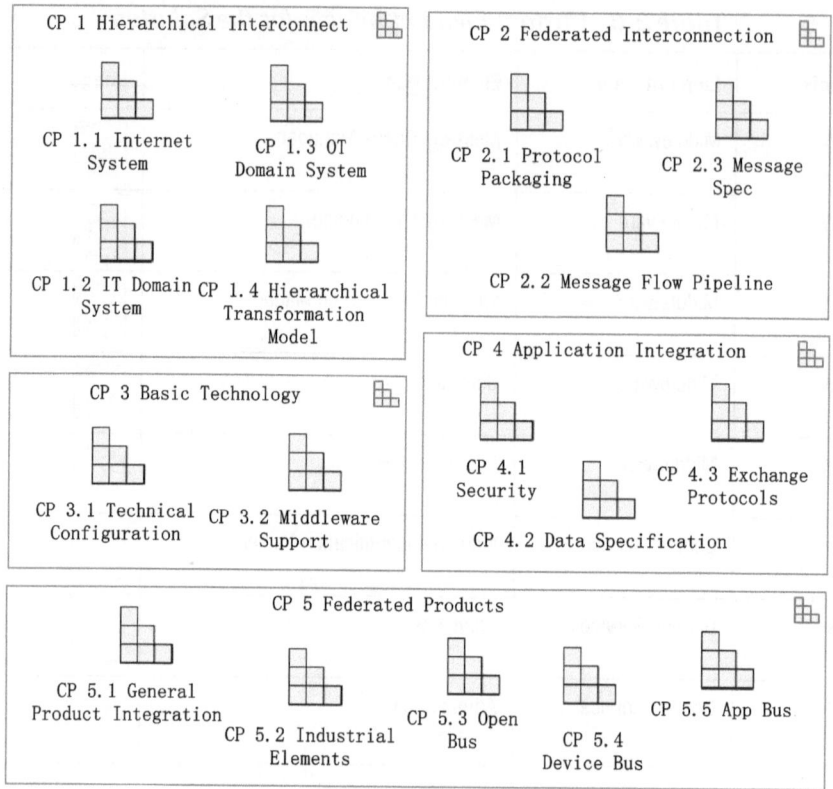

Figure 2-19. *EIS capability view, level 1*

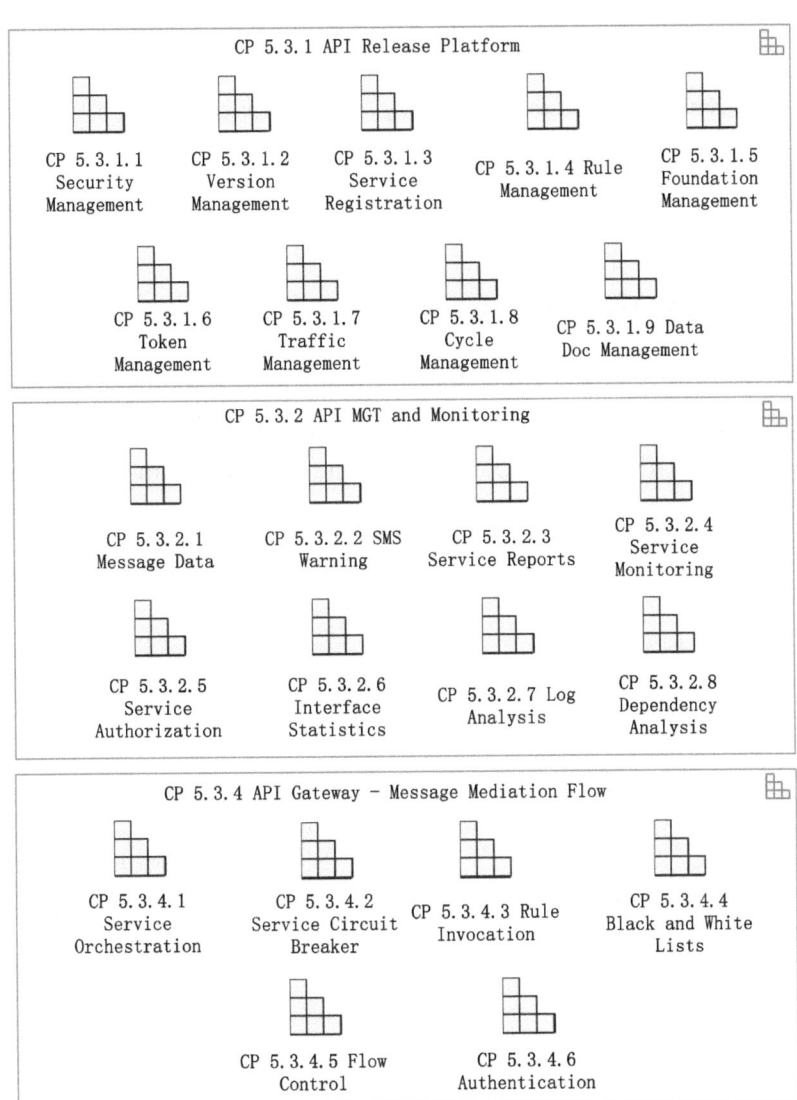

Figure 2-20. *EIS capability view, level 3*

HINT

Element Rendering Style: Practically, the capability view is better presented as a tabular form or tree view form for an overall observation. However, in ESA, key capabilities can also be presented as element nodes for better correlation with their resulting services or implementations.

Case Scenario Views

Figure 2-21 and Table 2-7 show the key quality requirements of the EIS solution. The *requirement tree* representation helps map the customer requirements critical to the solution and categorize the non-functional requirements for targeted architecture.

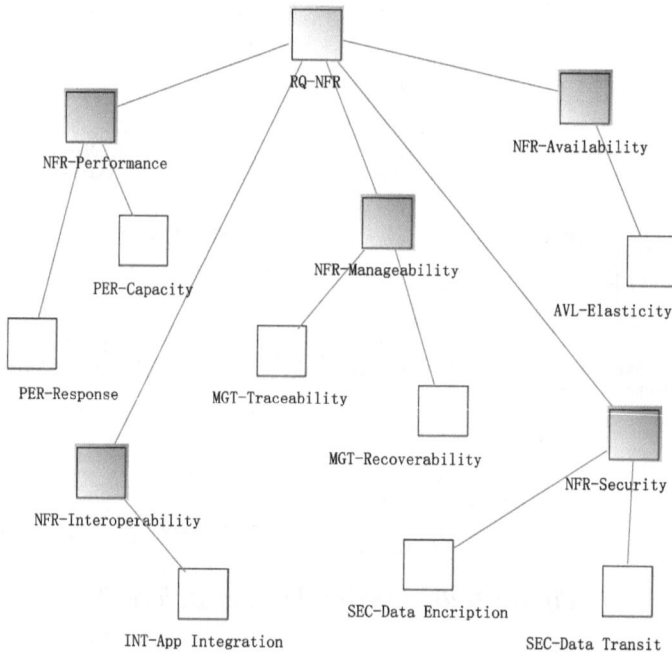

Figure 2-21. *EIS quality requirement tree*

Table 2-7. *Quality Requirement Tree of EIS Solution*

NFR Group	NFR Sub-Group	NFR Statement
NFR-Performance	PER-Response	NF-3 Intranet control within five seconds
		NF-4 the external network is controlled within ten seconds
		NF-5 the maximum range control time is within 60 seconds
	PER-Capacity	NF-2 support at least five concurrent information exchanges per second
		NF-6 the packet size is limited to 5MB
NFR-Manageability	MGT-Traceability	NF-9 keep the logs of data storage for no less than six months
	MGT-Recoverability	NF-11 provide error compensation for the resolution of the call processing calculation logic
NFR-Security	SEC-Data Encryption	NF-13 the ability to authenticate identity
		NF-14 the ability to achieve permission control
		NF-15 the ability to encrypt the transmission of the original message
		NF-16 the receiver can decrypt the original packet after it has been received
	SEC-Data Transit	NF-20 servers share the ports specified by them through the whitelist policy
		NF-21 to ensure the security of peer-to-peer access between servers
		NF-22 encrypt sensitive data during transmission
		NF-25 corresponds to the level of critical functionality served by the relevant key

(continued)

Table 2-7. (*continued*)

NFR Group	NFR Sub-Group	NFR Statement
NFR-Interoperability	INT-App Integration	NF-27 application system protocol encapsulation with IT domain
		NF-19 standardized protocol content packaging through the interface of heterogeneous system protocols
		NF-7 peer-to-peer exchange between heterogeneous systems
		NF-1 all domestic application systems should uniformly use UTF-8 for character set encoding
NFR-Availability	AVL-Elasticity	NF-8 use the buffering mechanism of message queues to reduce the performance overhead in the case of large concurrency
		NF-29 access control for the number of services
		NF-28 limit on the frequency of service calls per second

Figure 2-22 shows one use case model view for the EIS solution. In a system use case specification, associated NFRs can be included for functional mapping to see how any of these use cases or model structure is (partially) affected.

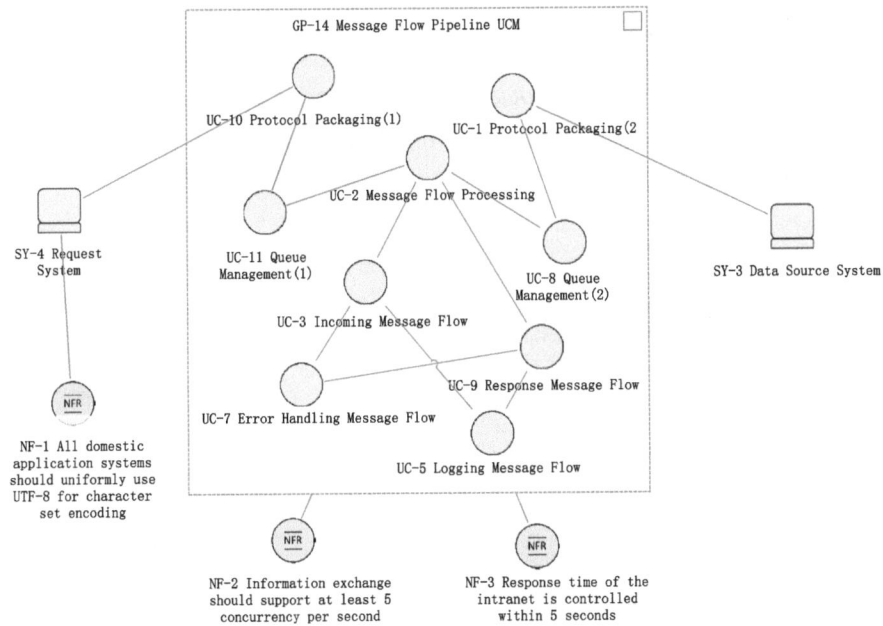

Figure 2-22. *An EIS use case model view*

Figure 2-23 shows a simple, self-explanatory process view of EIS.

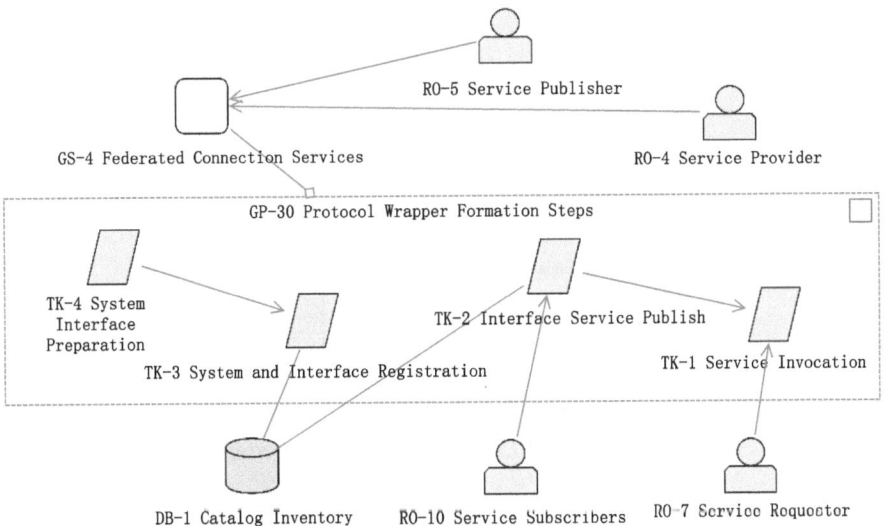

Figure 2-23. *An EIS process view*

Outline Views

Figure 2-24 shows an initial *block view* of enterprise integration service platform architecture outlining governing ideas, high-level building blocks, and layering.

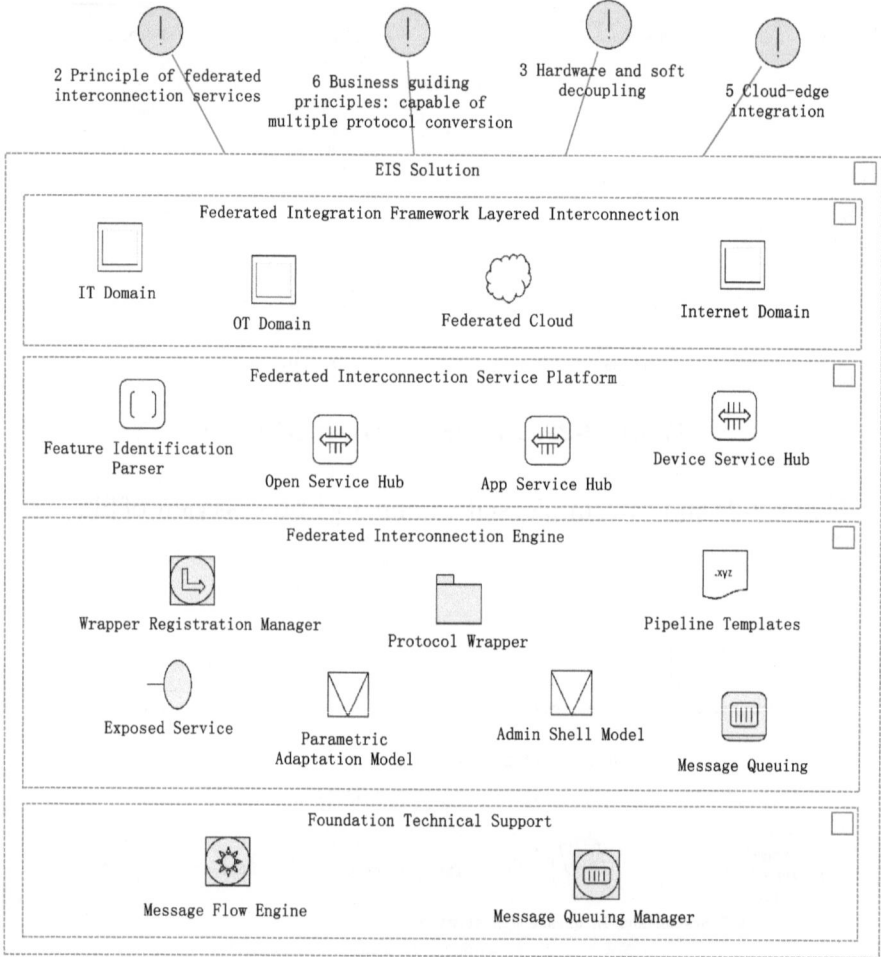

Figure 2-24. *EIS architecture block view*

Figure 2-25 demonstrates the EIS view frame architecture. This is a layered approach to expressing a large or complex architectural view in multiple views, with the top view containing *view frames* (VFs) that can be drilled down as child views.

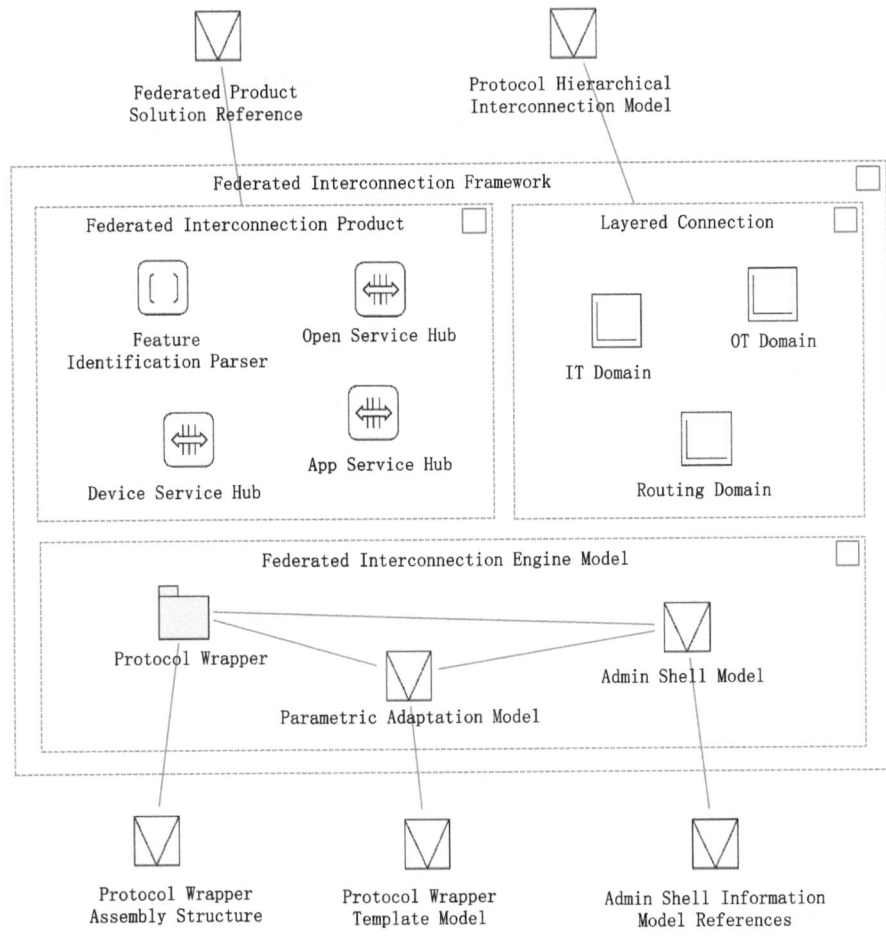

Figure 2-25. *EIS view frame architecture*

Metrics Views

This solution goes beyond just the solution environment. It also touches the enterprise and external environments as a reference architecture for wider use, so it requires metrics mapping to consider generality and reusability.

Figure 2-26 is an architectural metrics view for the EIS platform solution. The view displays some major metrics elements and their associations.

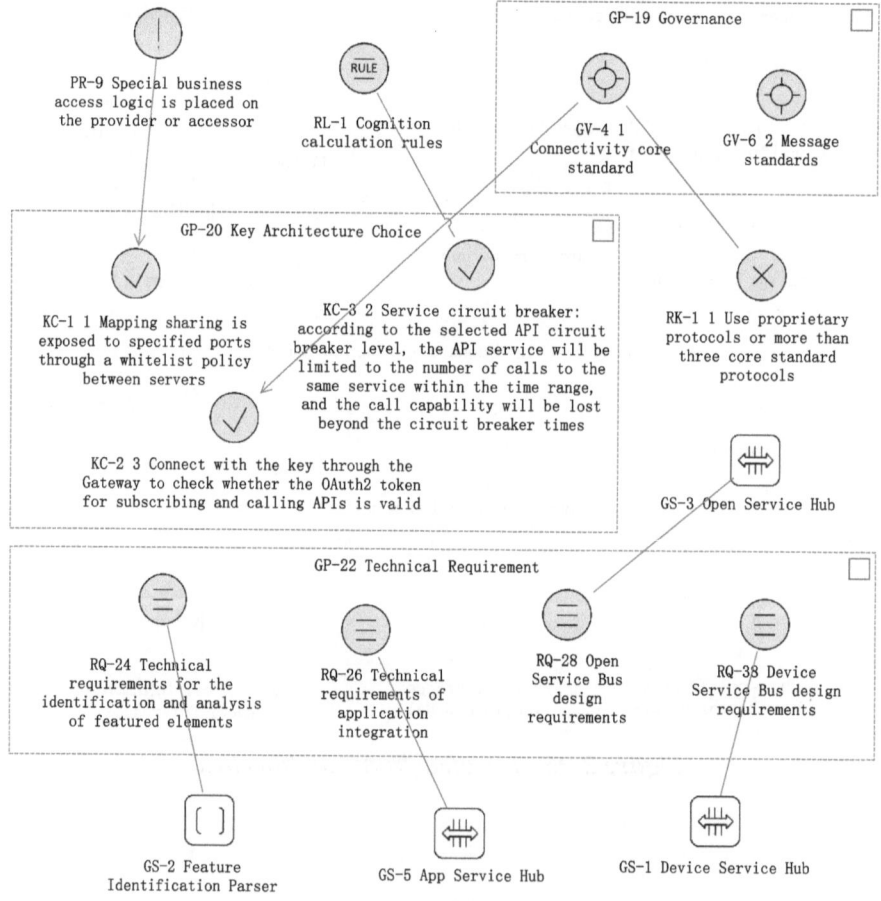

Figure 2-26. *EIS metrics view*

Metrics view is a must for any medium- to large-scale solutions. It allows a solution architect to think holistically about the key correlations between the five A-ESA metrics elements and their relevance to the solution elements, both functional and operational.

The metrics view can take different formats from a particular perspective and for a particular audience. Metrics elements can also be shown in other views where solution elements are the central pieces. Figure 2-27 is such an example and demonstrates an architecture outline view with requirements metrics mapping.

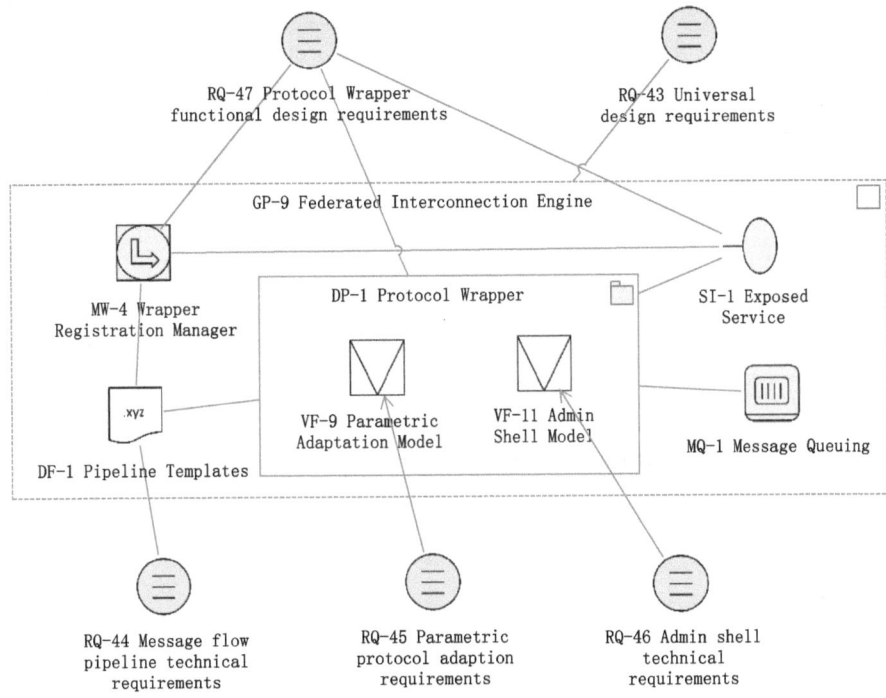

Figure 2-27. *Interconnection engine requirement mapping*

Pattern Views

There are many pattern views for this solution. Figure 2-28 is an *admin shell construct* pattern, the key architectural design in the EIS platform solution. This pattern view is at a certain level of abstraction and serves as a blueprint for the admin shell design. Table 2-8 shows some example properties of the EIS admin shell construct.

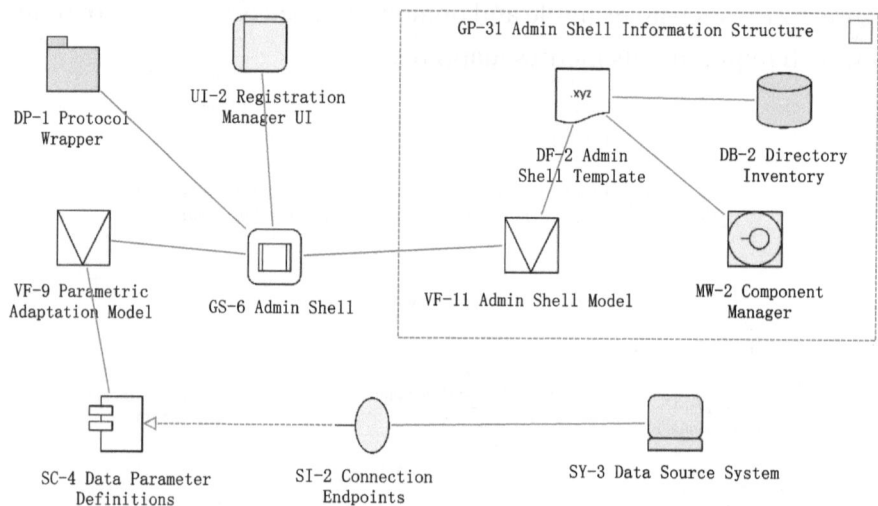

Figure 2-28. *EIS admin shell construct*

Table 2-8. *Example Properties of EIS Admin Shell Construct*

Element	Property Attribute	Property Description
VF-11 Admin Shell Model	Route Header	Record of the shell's own identity, asset/service identity, translation protocol, routing information
	Route Body - Operating Action	Operation instructions of this request, such as adding, querying, modifying, and deleting data and other operation functions
	Route Body - Business Data	Data content of this exchange, including input parameters and return parameters
	Target ID	Unique identifier of the target
	Catalog Inventory	Unified identity of the catalog inventory
	Component Manager	Globally unified identity for component management views that link to an external component management view
	Submodel="Type"	A submodel of "type" uniquely identified

(*continued*)

Table 2-8. (*continued*)

Element	Property Attribute	Property Description
DP-1 Protocol Wrapper	Wrapper unique identification ID	Unique ID that identifies the serviced interface automatically generated according to the service coding rules
	System identification ID	Identity of the system or device from the income service
	Message uniquely identification ID	Unique message ID dynamically generated at invocation time
	Interface connection parameters	Recording of the connection parameters of the interface, including IP address, username, password, and digital certificate
	Interface access	Access permissions, blacklists, and whitelists of interfaces
	Interface security authentication parameters	Interface access to digital certificates, etc.
	Interface protocol and matching adapter	Protocol type used by the interface and the corresponding protocol adapter template
	Interface inbound parameters	Input parameters for the interface call
	Interface outbound parameters	Return parameter of the interface call
	API message invocation template	XML message template
	Interface test packets	Packets for interface connectivity tests

Functional Service Views

Figure 2-29 shows a structural view of the EIS platform solution. It shows high-level relationships between the core EIS services, including functional services, middleware services, and solution guidance. Table 2-9 illustrates a portion of the properties from this view.

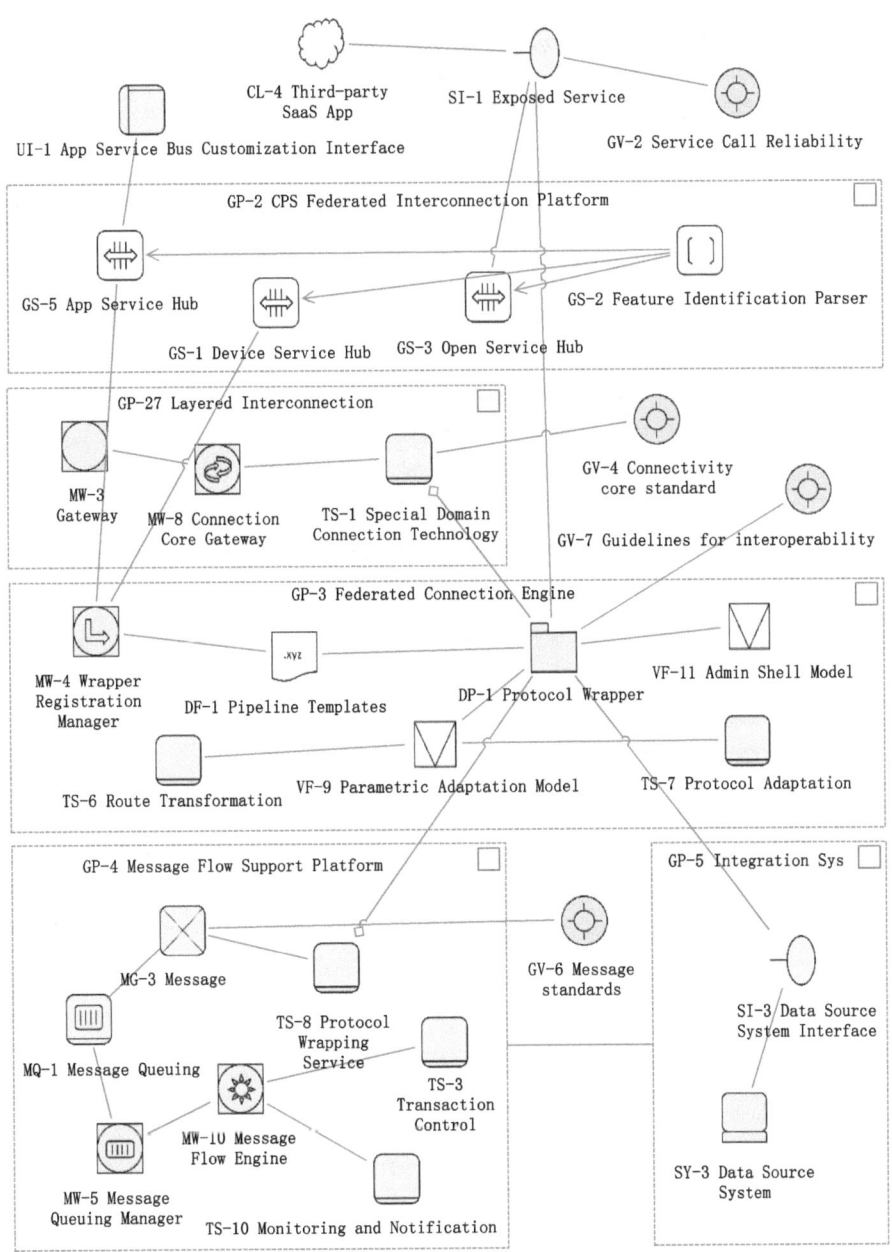

Figure 2-29. EIS structural view

Table 2-9. Part of Properties in Figure 2-29

Element	Property	Description
GV-6 Message Standard	Formula for calculating the amount of data contained in the business data in the message	$$S = \frac{\sum_{i=1}^{N} D_i}{1024^*1024} \quad \dots(1)$$ Note: S: The size of the data volume of each data recorded in the business data, unit: MB. N: The number of fields contained in each data record in the business data, unit: pcs. Max: Maximum value of field length in data record, unit: Byte. If the number of business data records contained in a transmission message is K, the unit is a message, and K must meet the following conditions: $$K \le \frac{1}{S} \quad \dots(2)$$
	Message size control range	5MB
	Message size not greater than	20MB

(*continued*)

Table 2-9. (*continued*)

Element	Property	Description

MG-3
Message
	Message Format	```<?xml version="1.0" encoding="UTF-8"?>```
		```<Service>```
		```<Route>```
		```<SourceSysID>02001</SourceSysID>```
		```<ServiceID>00010000000100</ServiceID>```
		```<SerialNO>2016051727635262123</SerialNO>```
		```<ServiceTime>20120621123000</ServiceTime>```
		```</Route>```
		```<Data>```
		```<Control/>```
		```<Request>```
		```<Dept>FICO</Dept>```
		```<EmployeeNo>12345678</EmployeeNo>```
		```</Request>```
		```<Response/>```
		```</Data>```
		```</Service>```

GV-2 Service Call Reliability	Reliability Checklist	1. Secure API communication if any 2. Ensure credential setup if any 3. Plan response caching 4. Set quotas and rate limits if any 5. Ensure a backup, restore and recover strategy 6. Configure multiple regions for cloud hosting 7. Avoid API chattiness

Figure 2-30 is a *service collaboration view* for the EIS platform solution. The protocol wrap-up service, one of the significant scenario cases, is a composite service that works as a step-by-step process. Table 2-10 shows an example of the use of properties in the EIS collaboration view.

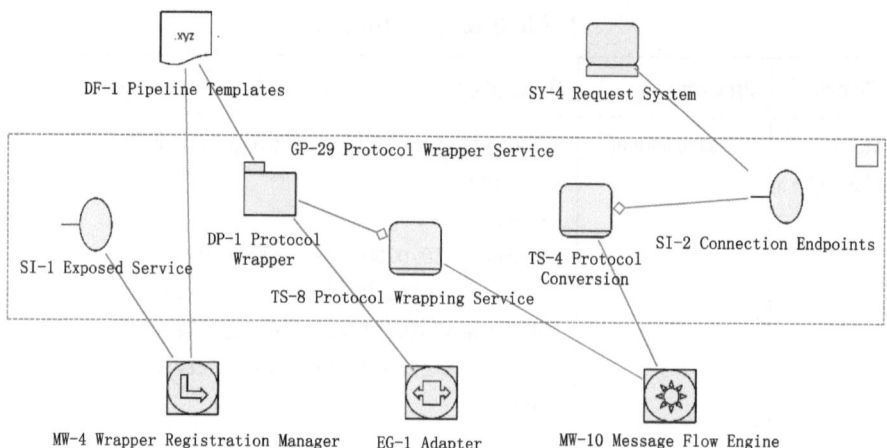

Figure 2-30. *An EIS service collaboration view*

Table 2-10. *Example Properties in Figure 2-30*

Element	Protocol Property	Property Specification
EG-1 Adapter	MQ-Message Queue	Address\Port\Queue Name\Username, etc.
	HTTP-Mainstream Web	URL\Parameters\Method\Username, etc.
	SOAP-WSDL Web Protocol	WSDL address\parameters, etc.
	FTP-File Transmission	Address\Port\username\Password\Directory, etc.
	RFC-ERP System Specific Protocol	Address\Port\Username\Password\Module \ Number\RFC Function Name, etc.
	MQTT-IoT Protocol	Address\Port\Subject, etc.
	OPC UA/OPC DA-Standard	Address\Port\Subject, etc.
	DB-Database Adapters	Address\Port\Username\Password, etc.
	Rest API-Lightweight RPC	Address\Parameters, etc.

Deployment Views

Figure 2-31 is a deployment view that highlights the relationships of
key functional services for the EIS *platform device service hub* in the
operational environment. This view is intended for considerations of
additional non-functional requirements, including rules. It is an initial
view for analysis and static testing.

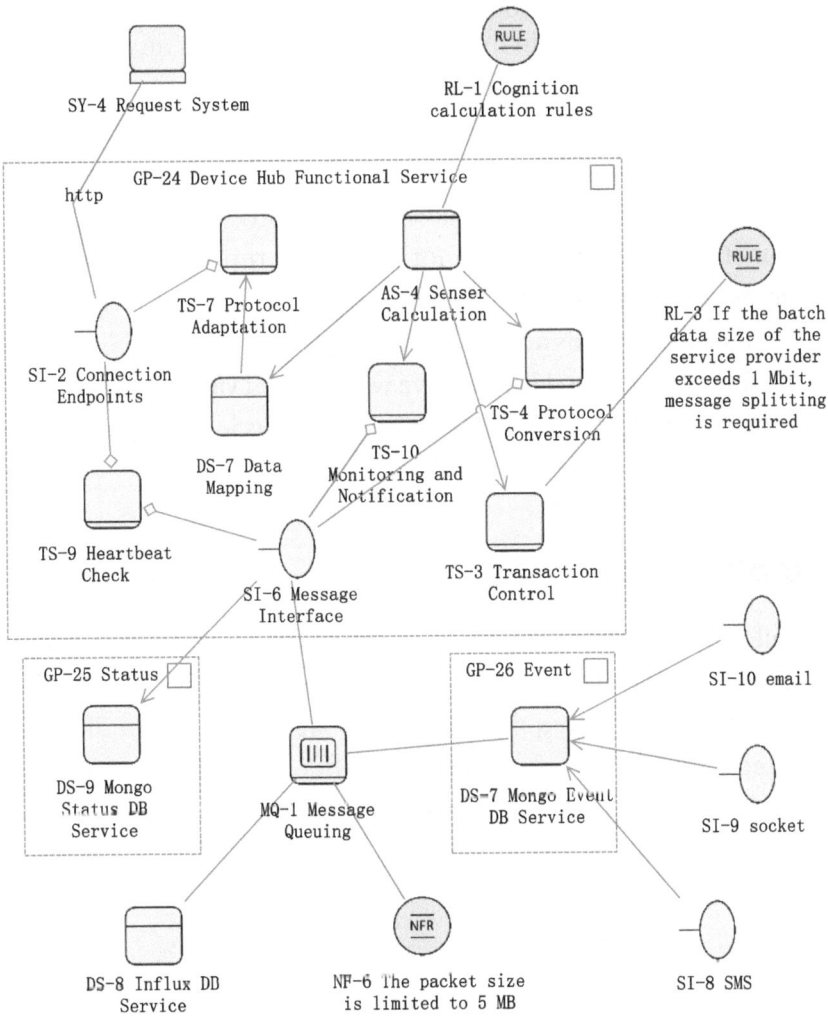

Figure 2-31. *EIS deployment view*

A-ESA Profile 5: Application Architecture

This solution is an *application governance architecture* in a *service-based architecture (SBA)* for the *order processing solution*, focusing on application modeling techniques and architectural design governance.

In general, it is more difficult to model an application architecture than other architectures, because business logic modules, if not architected for flexibility, are harder to change after they are deployed in the production environment. So, keep this in mind: the *cost of change* for an application architecture is much higher than for a technical architecture.

When creating functional views, consider these proven architectural techniques: *high cohesion, loose coupling, proper granularity, flexible modularity, isolation, proper dependency level,* and *better layering.* Use A-ESA functional service elements for intuitive partitioning of service boundaries and characteristics. Iteratively refine architecture views to meet both business needs and future changes.

Note that the solution is illustrative with limited views for easy and clear understanding of selected architectural points including *modularity, granularity,* and *flexibility.*

Case Scenario Views

In an Agile project environment, a *user story*, as you know it, is an informal form of use case. User stories use non-technical languages, and they are in fact not software system requirements. They are ultimately handled from an architectural solution perspective, with added system requirements. Table 2-11 shows a *register customer* user story property.

Table 2-11. *User Story Property: Register Customer*

Story Item	Statement
Statement	As a customer representative, I would like to help register customers, so that I can provide convenience to new customers who are in need of assistance.
Task	– Get postal code for delivery area – Get customer profile information including name, address, and email – Provide a unique reference number to the customer
Time	0.5 Day
Complexity	XS T-shirt size
Development By	Julie Wang
Acceptance Criteria	Functional Service SLAs

Architecture Outline Views

Figure 2-32 and Table 2-12 show an architecture block view that outlines the application layers. The solution is an IT service-based architecture that strikes a balance between domain subdivision and technical isolation. The view here is a typical architecture layer partition.

Table 2-12. *AOV-2 Architecture Outline View Properties*

View	Property	Description
AOV-2 Architecture Outline	Type	Architecture Overview - Layer
	Style	IT Service-based Architecture
	IsGoverningView	Y
	SystemContextDomain	OPS Boundary
	ArchState	To-Be
	SolutionProject	OPS Application Governance Solution

HINT

Domain Partitioning vs. Layering: Domain partitioning is an obvious grouping by domain purpose from an application service architecture perspective, while layering is generally an architectural partitioning from high business level to lower technical level. Note that the concept of partitioning may mean different things to different solution architects. The application of the concept in the modeling is the best way to tell what it really means.

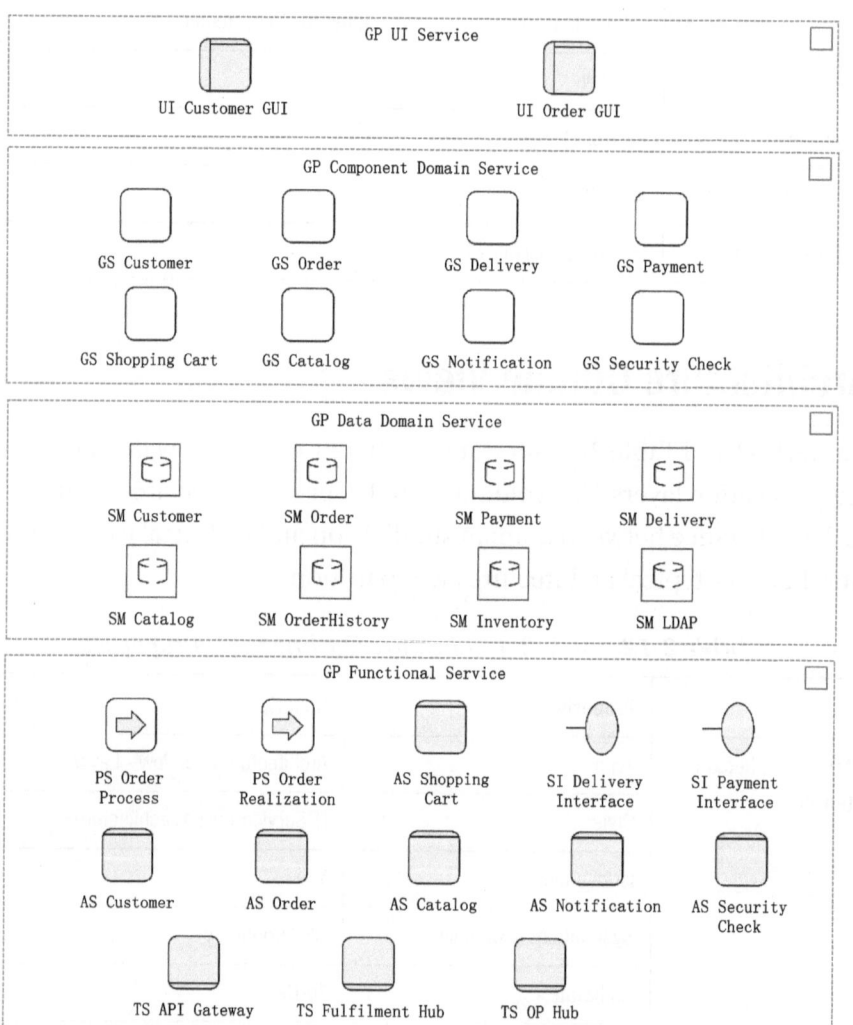

Figure 2-32. *AOV-2 architecture outline view*

Metrics Views

Figure 2-33 shows the important governance metrics for the solution. GV-1 serves as the major guidance, referred to as the guiding governance. This is the key architecture view of the application architecture. It specifies critical architectural governances and embodies architectural thinking on how to construct an application architecture that is subject to change.

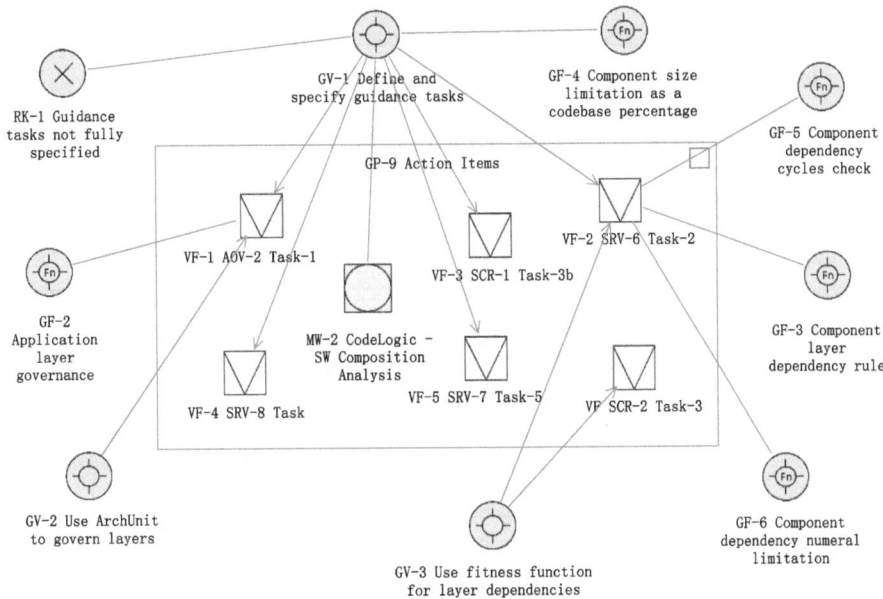

Figure 2-33. *MTS-1 metrics view*

When rendering metrics elements on the Figure 2-33, the general steps are as follows:

1. Place the application architecture governance elements.

2. Track all architecture views that are subject to governance.

3. Consider governance functions (either directly or indirectly applicable to architecture views or development design).

4. Validate associated metrics elements in an iterative process.

Table 2-13 shows the architecture element properties of the metrics view. Depending on the supporting solution architecture tool, the same content can be expressed in different formats, and Table 2-14 shows the GV-1 element property in a simpler name-value pair.

Table 2-13. *GV-1 Element Properties of the Metrics View*

Element	Step	Task	View	Element
GV-1 Define and specify guidance tasks	1	Define component layers	AO-2	GF-2
	2	Define component domains	SRV-6	GF-3
	3	Define component structural relationship	SCR-2	
	3a	Determine granularity		GF-4
	3b	Determine cohesion	SCR-1	
	3c	Determine component dependency	SRV-6	GF-5, GF-6
	3d	Detect common code base		MW-2
	4	Define domain services	SRV-8	
	5	Define functional services and relationships	SRV-7	

Table 2-14. *Simplified Expression of Table 2-13*

Element	Property Name	A-ESA Value
GV-1 Define and specify guidance tasks	1. Define component layers	AOV-2 I GF-2
	2. Define component domains	SRV-6 I GF-3
	3. Define component structural relationship	SCR-2
	3a. Determine granularity	GF-4
	3b. Determine cohesion	SCR-1
	3c. Determine component dependency	SRV-6 I GF-5 I GF-6
	3d. Detect code base logic (bottom-up approach)	MW-2
	4. Define domain services	SRV-8
	5. Define functional services and relationships	SRV-7

As seen in Table 2-14, the top-level governance element GV-1 defines a group of tasks, each of which is delegated to an architectural artifact (either a view or an element). Each view frame (VF) element corresponds to an architecture view.

Note that the governance function (GF) elements specify architectural governance rules (architectural function, architectural design quality metrics, or cross-cutting concerns) using pseudocode or language-specific rules.[8] In addition, architectural governance can be supported by middleware tools such as MW-2 CodeLogic (software composition analysis tools), as shown in Figure 2-33. Table 2-15 provides a list of the governance functions shown in Figure 2-33.

---

[8] Supported from tooling vendors such as ArchUnit, NetArchTest, and JDepend.

Table 2-15. *Element Properties of Governance Functions.*

Element	Governance Function
GF-2 Application layer governance mapping ArchUnit specification	**GF-2 Governance Function Specification:** – Controllers layer may not be accessed by any layer – Services layer may only be accessed by Controllers layer – Repositories layer may only be accessed by Services layer

	GF-2 ArchUnit Code Specification:
	ArchUnit Code Specification
	`@ArchTest`
	`static final ArchRule layer_dependencies_are_respected =`
	`layeredArchitecture()`
	
	`.layer("Controllers").definedBy("com.mscharhag.archunit.`
	`layers.controller..")`
	`.layer("Services").definedBy("com.mscharhag.archunit.`
	`layers.service..")`
	`.layer("Repositories").definedBy("com.mscharhag.archunit.`
	`layers.repository..")`
	`.whereLayer("Controllers").mayNotBeAccessedByAnyLayer()`
	`.whereLayer("Services").mayOnlyBeAccessedByLayers("Control`
	`lers")`
	`.whereLayer("Repositories").mayOnlyBeAccessedByLayers("Ser`
	`vices");`

GF-3 Component layer dependency rule mapping NetArchTest specification	**GF-3 Governance Function Specification:** – Types in current domain that reside in namespace should – Have dependency on API namespace – Have dependency on Infrastructure namespace

	GF-3 NetArchTest Code Specification:
	`Types.InCurrentDomain()`
	`.That().ResideInNamespace(DomainNamespace)`
	`.Should().HaveDependencyOn(ApiNamespace)`
	`.And().HaveDependencyOn(InfrastructureNamespace)`
	`.GetResult().IsSuccessful`
	`.Should().BeTrue();`

(continued)

Table 2-15. (*continued*)

Element	Governance Function
GF-4 Component size limitation as a codebase percentage mapping pseudocode	**GF-4 Governance Function Specification:** 1. Loop through the codebase directory 2. Get the directory namespace structure 3. Identify and verify all components 4. Step through the source code to get the total statements 5. Loop through the component list and divide each component's statements by the total statements 6. If the result is greater than 12%, flag it 7. Show the flagged components (if any)
	GF-4 Governance Pseudocode: *FOR each directory IN codebase directory* 　　　　*STORE directory IN namespace structure* *FOR each component IN component list* 　　　　*RETRIEVE statements IN each component* 　　　　*ADD total statements* *FOR each component IN component list* 　　　　*DIVIDE each component's statements BY total statements* 　　　　*FLAG component IF result > 12%* *GET a list of flagged components (if any)*

(*continued*)

Table 2-15. (*continued*)

Element	Governance Function
GF-5 Component dependency cycles check mapping JDepend specification	**GF-5 Governance Function Specification:** Check all packages for cyclic dependencies

	GF-5 JDepend Code Specification:
	
	``java
	public void checkAllPackagesCycles()
	{
	  Collection packages = jdepend.analyze();
	  if (jdepend.containsCycles())
	  {
	    for (Iterator i = jdepend.getPackages().iterator();
	    i.hasNext();)
	    {
	      JavaPackage jPackage = (JavaPackage) i.next();
	      if (jPackage.containsCycle())
	      {
	        System.out.println("Cycle at " + jPackage.getName());
	      }
	    }
	  }
	  assertEquals("Cycles exist", false, jdepend.
	containsCycles());
	}
	``

(*continued*)

Table 2-15. (*continued*)

Element	Governance Function
GF-6 Component dependency numeral limitation	**GF-6 Governance Function Specification:** 1. Loop through the codebase directory and get a list of components and their source code 2. Count all incoming and outgoing references for each component 3. List all components with a reference count greater than 10

Note:
- The governance function rules prefer to use pseudocode or optionally sample code in conjunction with a more generic metrics tool such as ArchUnit and JDepend.
- There are also software architecture design tools (such as FindBugs, Structure101, Eclipse Metrics, Lattix, and PMD Source Code Analyzer) that can discover and enforce metrics from the source code level for visibility and analysis. For example, Lattix (applicable to Java, C++, and .Net) can handle metrics such as cyclomatic complexity, interface count, average dependency, coupling, system stability, inter-component cyclicity, path complexity, weighted impact, etc.

HINT

Role of A-ESA Governance Function: A-ESA's governance function does not tie to any specific technology, although A-ESA model tool will provide a mapping transition between model governance specifications and tool detection and enforcement. ESA goes beyond software architecture and addresses functional and operational aspects at a less granular level. In the case of green-field application architecture, governance functions will initially serve only to provide guidance. In contrast, for brown-field architectural work or validation, governance functions will generally enforce architectural conformance.

Pattern Views

Pattern view is a powerful means of expressing application architectural solutions, especially governance functions that touch on large-scale and distributed cross-cutting concerns. The view can be leveraged from proven solutions or can be reused from similar solutions. The view can then be realized as a technical service, a microservice, a middleware service, a mid-platform service, or a cloud service.

Figure 2-34 shows a *TCC[9] transaction* view that is considered in the order process solution. This type of transaction is widely used in enterprise solution transactions such as ERP integration scenarios. The caveat with TCC is that each scenario requires mapping and coding of its transaction logic, so its reusability is a common issue.

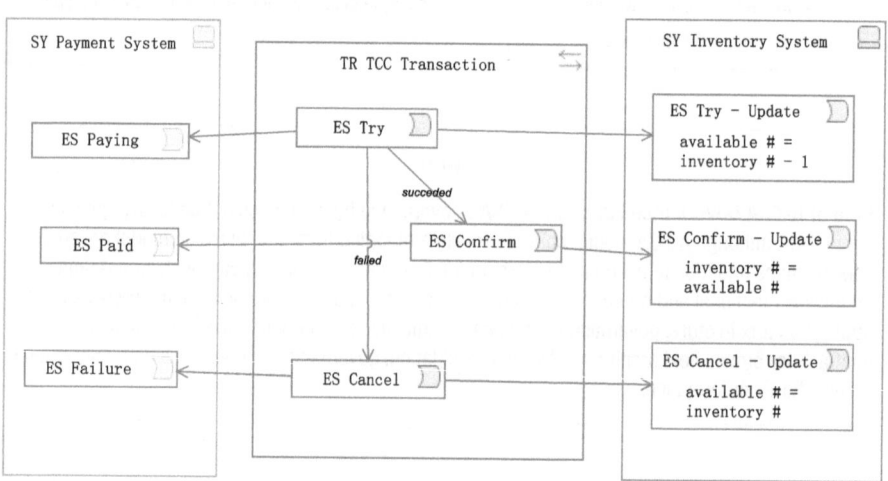

Figure 2-34. *TCC distributed transaction*

Figure 2-35 shows another *distributed transaction* pattern in a relatively weak consistency approach as a key architectural consideration. It's an assured asynchronous message communication.

---

[9] Try, Confirm, Cancel

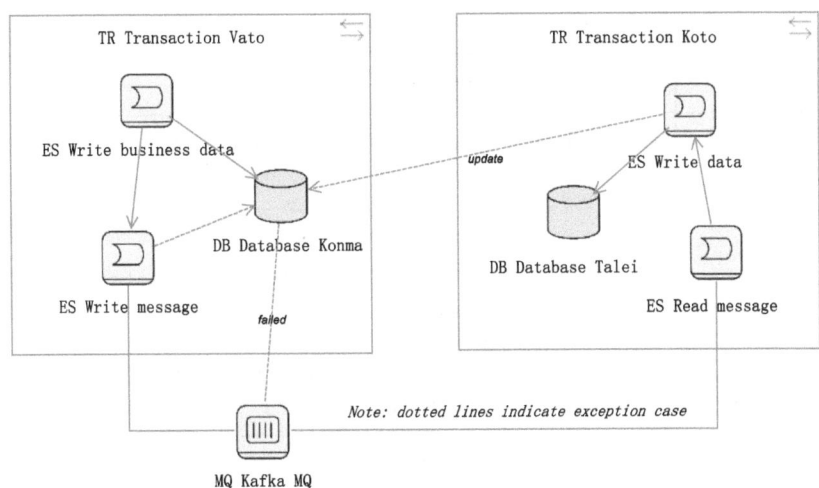

Figure 2-35. *Message-based distributed transaction*

Figure 2-36 exhibits an *idempotent consumer* pattern. It de-duplicates incoming messages by detecting duplicate messages by identity. Table 2-16 is the architecture decision for the idempotent implementation.

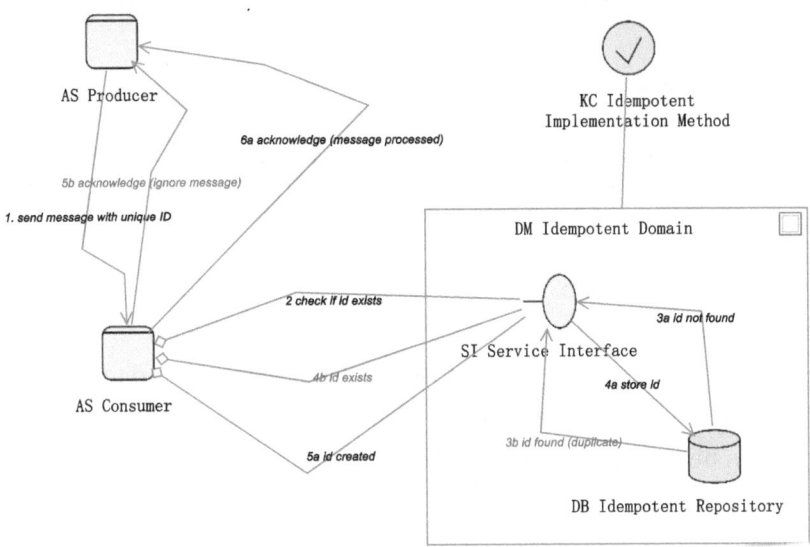

Figure 2-36. *Idempotent consumer pattern*

Table 2-16. *Key Consideration on Idempotent Implementation*

Element	Property	Description
KC-4 Idempotent Implementation Method	Alternative	Restful API interface limitation: Not all operations ensure idempotence.
		Database optimistic locking limitation: Additional fields must be added to the corresponding business table in the database.
		Database's unique primary key requirement: A globally unique primary key ID must be generated.
		Anti-duplication token restrictions: A globally unique token string must be generated, and the third-party component Redis must be used for data validation.
		Downstream unique serial number usage restrictions: The third party must pass the unique serial number; the third party component Redis is required for data validation.
	Decision	Use database unique primary key to achieve idempotence.

Note:
- An idempotent operation is an operation for which the effect of performing the same operation multiple times is the same as performing it once.
- Many messaging systems, message brokers, and service buses have capabilities to help eliminate duplicate messages, and some also provide exactly-once semantics.

Figure 2-37 shows an *SOA service framework* pattern used for distributed service communication via RPC.[10] It is a high-level architectural pattern view that exposes interfaces in selected technical services while hiding information in other technical services. The design pattern can be further detailed by the development team.

---

[10] Remote Procedure Call

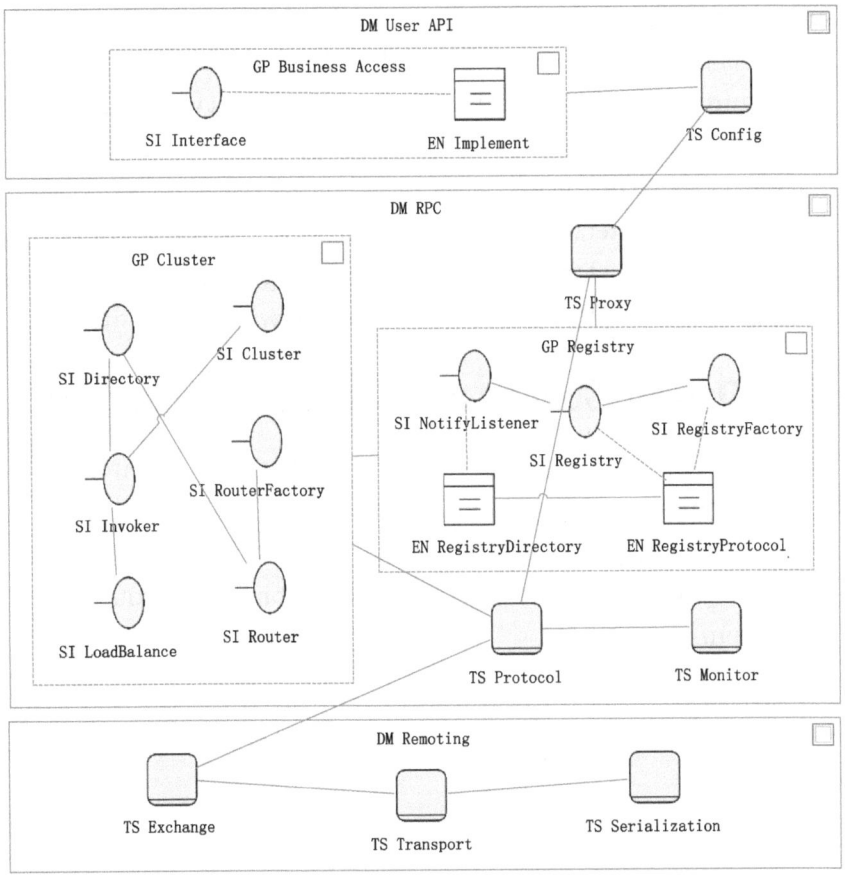

Figure 2-37. *SOA service communication framework*

Functional Views

Functional views are the focus of application architecture. This level of architectural design is the most critical and challenging part of solution architecture, as it has a major impact on architectural *adaptability* or *changeability*.

Functional services fulfill the enterprise plans, case scenarios, and stakeholder requirements, as well as the transitional responsibilities to the operational environment. Thus, the functional service view is closely related to business architecture, data architecture and technical architecture. Each service needs to consider the principles of cohesion, coupling, and isolation. What also matters more in this view is layer partitioning, generic domain definition, and granularity determination, which is an iterative process.

Application architecture, much of which is beyond the AI automation capabilities, requires a lot of architectural thinking to achieve architectural conformance, as partially exemplified in the following functional views.

Generic Domain Context and Service Specification

Figure 2-38 shows an *SOA generic domain service* view that better reflects the domain interaction relationship in a selected profile. *Generic service* (GS) element is used initially from a business perspective and could be morphed into *domain* (DM) element for more rigorous design consideration.

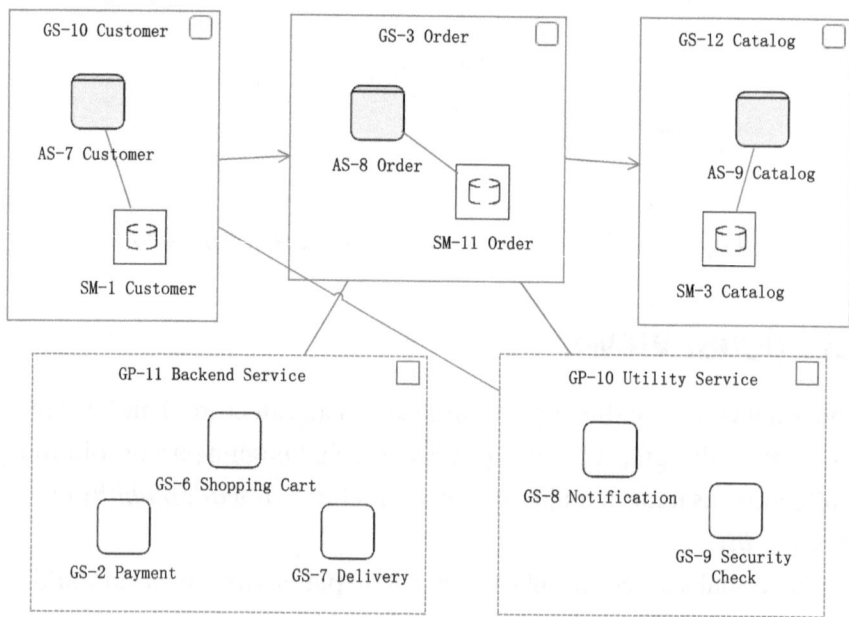

Figure 2-38. *SRV-8 generic domain service*

Table 2-17 shows an initial *draft* of the generic domain service analysis in tabular form and Figure 2-39 shows its application *service relationship* view (the core functional view) in the mobile application solution. Note that the generic domain services act as a liaison to the component domain services in the next section.

Table 2-17. *An Initial Generic Domain Service Identification*

Generic Domain Service	Legacy Service	Functional Service
Customer		Customer GUI
	Customer	
Order		Order GUI
	Order	
		Order Realization
Shopping Cart		Shopping Cart
Payment		Payment Interface
Catalog	Catalog	
Delivery		Delivery Interface
Security Check	Security Check	
Notification	Notification	
Integration		Order Processing Hub
		Data Sync & Replica
		API Gateway
		Fulfillment Hub

Note:

– There are other considerations such as mobile and web application differentiations.

– The order process can be architected as an application service if it's implemented as a control logic, as a business process if it's handled by a business process engine, or as part of a microservice if it's justified by an architectural decision.

Figure 2-39 demonstrates typical functional elements with different characteristics for a high-cohesion and loose-coupling structure. The *order process* (PS-1) performs control logic (or orchestration, if warranted). The *order realization* (PS-2) serves as a composite service performing service choreography. Various *application logic services* (ASs) perform their atomic business-dependent functionalities. The *technical service hub* (TS-2) provides asynchronous communication and distribution. And *service interfaces* (SIs) specify contracts for communication protocol and loose coupling.

The application logic services in Figure 2-39 could have been specified differently as *data services* if the business logic handling is done at the process level. Or *a process service* could access database directly without an application logic service or a data service. These could all be valid choices, depending on the business scale, SLRs, and architecture decision(s). In a word, whatever the choice is, it should at least meet the current requirements of the entire solution, and optimally meet future requirements.

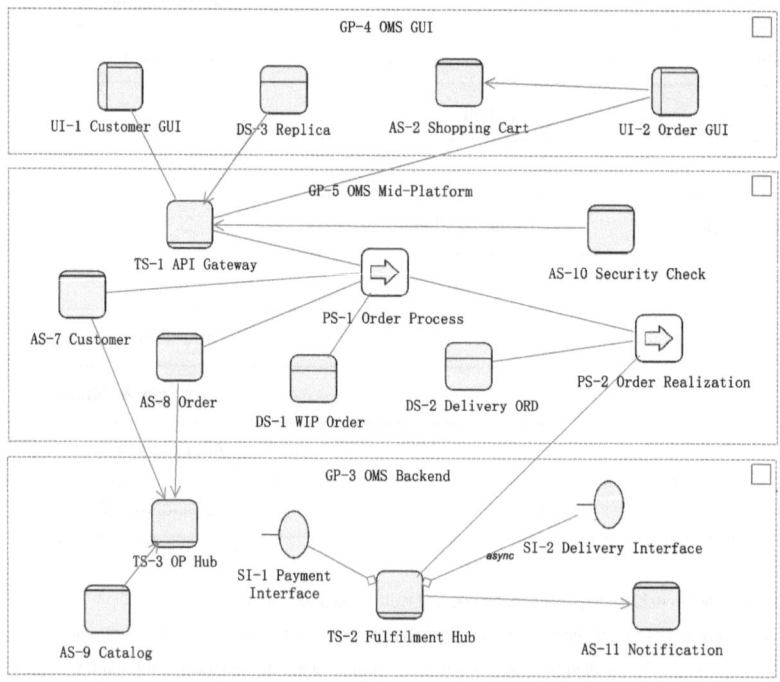

Figure 2-39. *SRV-7 service relationship*

HINT

Tabular Form vs. Relationship View: Tabular form can be an easy way to do the initial analysis of architecture structure, but the architectural view better reflects element relationships and element properties. What is more, it allows for dynamic relationships in an interactive scenario or walkthrough simulation. It also enforces holistic contemplations and modeling integrity. For example, imagine how mobile data synchronization is portrayed in the view and how workflow compensation (if any) is handled.

Table 2-18 shows some of the element properties in Figure 2-39. These element properties, part of a meaningful architectural work, serve as a guide for further design work.

Table 2-18. *Functional Element Properties in Figure 2-39*

Element	Property	Description
UI-1 Customer GUI	Name	Mobile Customer UI
	AccessFrequency	Medium
	InterfaceSystem	Mobile Phone Device
	DesignDoc	OPS GUI Design Document.doc
	DesignTool	Axure UI I MAUI
	GUIDesignStyle	Responsive
	TestSpec	Wireframe Test IV
	UserType	Naïve
	Concurrency	Medium

(continued)

Table 2-18. (*continued*)

Element	Property	Description
AS-2 Shopping Cart	Name	Mobile Shopping Cart
	Type	Application Logic Service
	TransactionRate	High
	ProgrammingLanguage	Java
	DevTools	IntelliJ IDEA
	HasDegradedFunction	N
	GranularityLevel	Composite
SI-2 Delivery Interface	Name	Order Delivery Interface
	Type	Service Component Interface
	TargetProtocol	jsm
	Scope	Private
	RequestMethod	GET
	InputParameter	Order Data Service
	ReturnParameter	Order Data Service
	TestingURL	/v3/delivery
TS-2 Fulfillment Hub	Name	Order Fulfillment Service
	MiddlewareAssociation	TBD
	TransactionType	Intermittent
	RealizationApproach	Service Hub Solution

(*continued*)

Table 2-18. (*continued*)

Element	Property	Description
DS-1 WIP Order	Name	WIP Order Data
	Type	Data Service
	DataFormat	Structured
	DataStore	Data Object
	IsEncrypted	N
	IsAudited	Y
	Volatility	High
	Concurrency	Medium
	UsageIntensity	High
	TransactionIntegrity	High
	DataSplit	OrderID

Figure 2-40 is a functional collaboration (interaction) view that shows the service interaction of *order placement and payment,* which a significant architecture flow that ensures high availability. Depending on the architectural focus, each interaction view may reflect only one perspective such as logical, physical, application, technical or development. The collaboration view enables a thinking process for a use case scenario walkthrough. For example, how the delivery slot information is represented and how it is handled when the backend delivery gateway experiences a network hiccup.

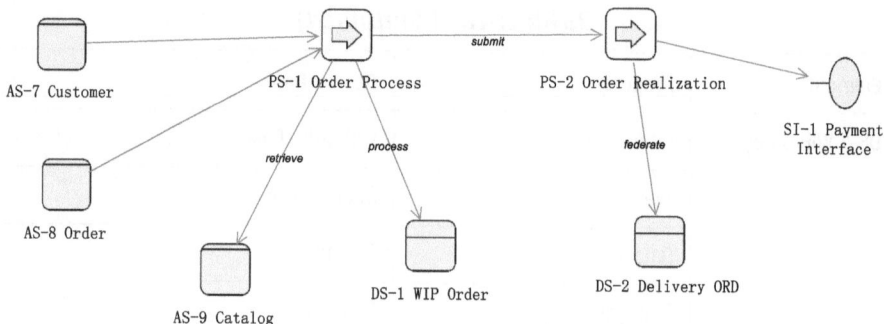

Figure 2-40. *SIV-1 order processing interaction*

HINT

Relationship vs. Interaction: At the solution architecture level, use relationship view (sunny day scenario) for functional structure and relationship to keep architecture uncluttered. Use interaction (or collaboration) view only to exemplify architectural considerations of significant use case(s) and critical error handling walkthroughs (rainy day or game day scenarios) such as order placement and payment.

Component Domain Service Structuring

Figure 2-41 shows an architecture view of *component* domains and dependencies. This view reflects the service component decomposition, which in turn determines the component modularity through layering, categorization, and service level characteristics. Component decomposition is an iterative and holistic process guided by the GV-1 element specification (see Figure 2-33).

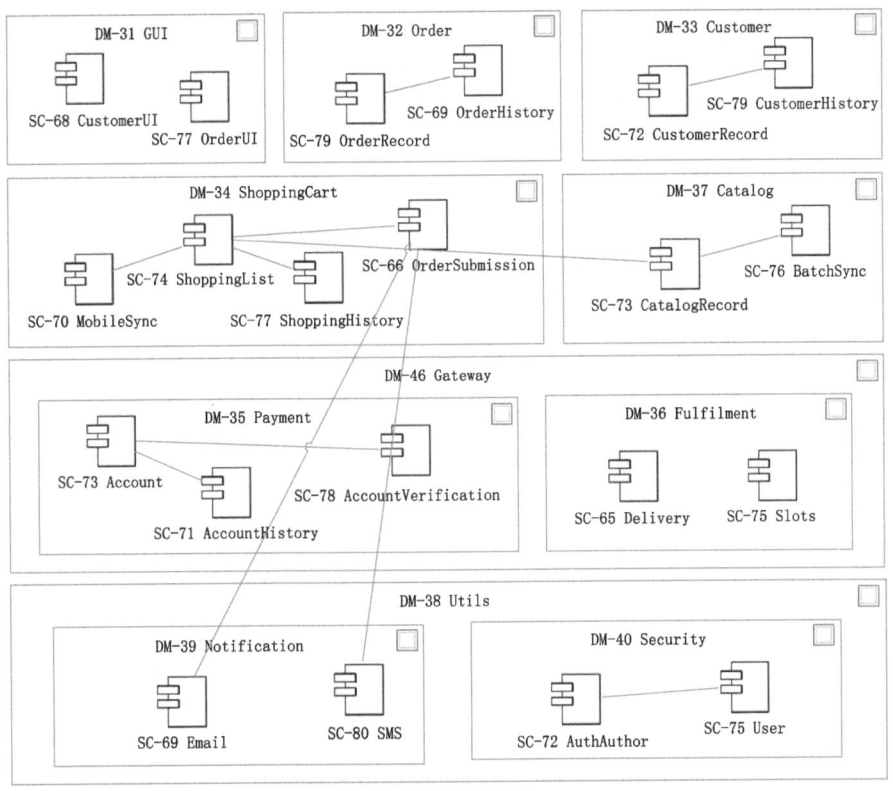

Figure 2-41. *SRV-6 component domain and dependencies*

When determining the component granularity and component cohesion, it is important to clearly define the *responsibilities* of each component, which are usually provided by the business or data domain experts. Figure 2-42 shows a pop-up view of the selected component responsibilities. Table 2-19 shows the component responsibilities in the form of architectural properties. Along with properly assigned component responsibilities comes component modularity with less tangled dependencies, thus less coupled architecture.

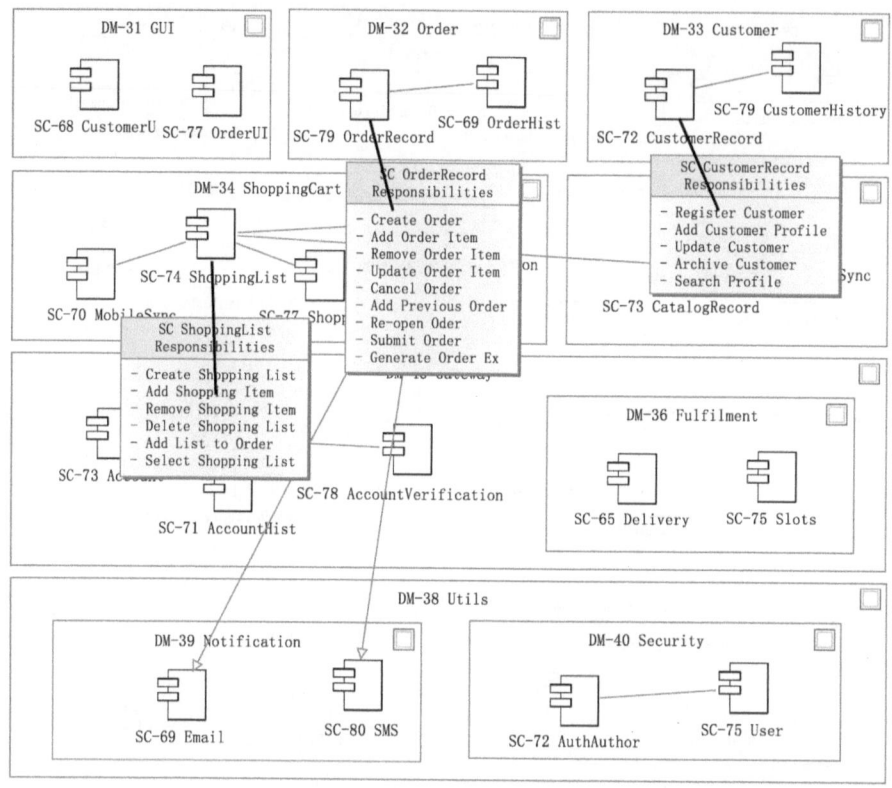

Figure 2-42. *SCR-1 view properties of component responsibilities*

Table 2-19. *SCR-1 Property of Component Responsibilities*

Domain	Service Component	Component Responsibilities
DM-32 Order	SC-79 OrderRecord	– Create Order – Add Order Item – Remove Order Item – Update Order Item – Cancel Order – Add Previous Order – Reopen Oder – Submit Order – Generate Order Exception
DM-33 Customer	SC-72 CustomerRecord	– Register Customer – Add Customer Profile – Update Customer – Archive Customer – Search Profile
DM-34 ShoppingCart	SC-74 ShoppingList	– Create Shopping List – Add Shopping Item – Remove Shopping Item – Delete Shopping List – Add List to Order – Select Shopping List

Figure 2-43 shows the component *namespace* view, another side view of Figure 2-41. Although the namespace specification is more at the design and development end, it is an effective means of linking architectural design and application landing.

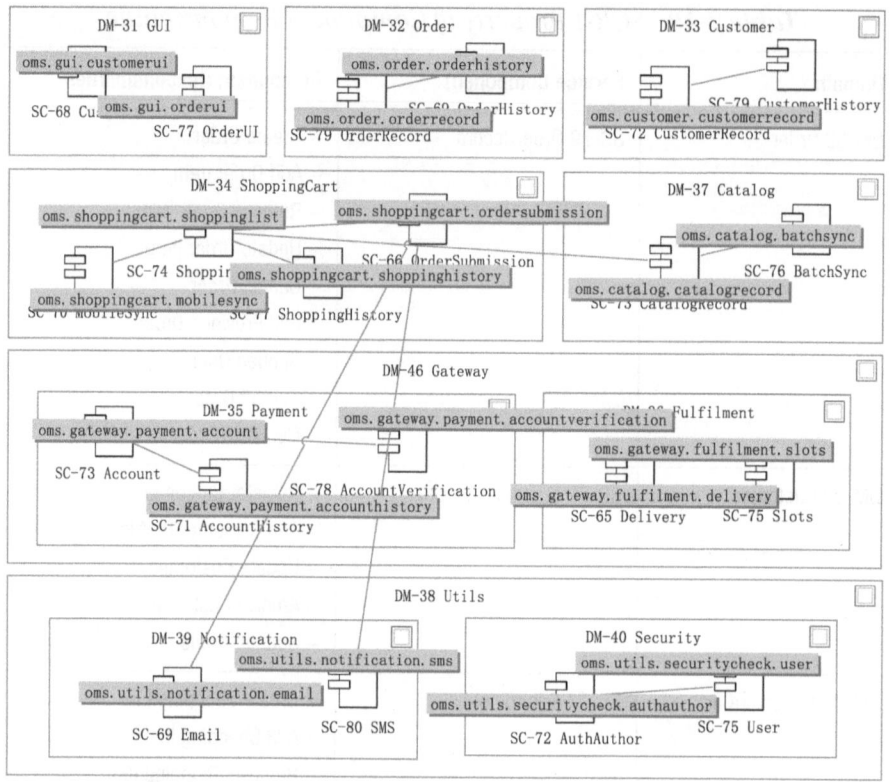

Figure 2-43. *SCR-2 component namespace structure*

The namespace can be in the form of a property attribute or grouped in a tabular or tree view form. The pop-up view in Figure 2-43 is a tooling support feature that facilitates architectural mapping and verification.

Figure 2-44 and Table 2-20 illustrate how requirements affect component modeling. RL-4 and RL-5 are business rule requirements. NF-5 and NF-6 are both architectural concerns. NF-5 affects solution capacity planning, availability, and performance, and NF-6 creates a throughput requirement. NF-3 and NF-2 are both easy to handle and don't require much architectural attention.

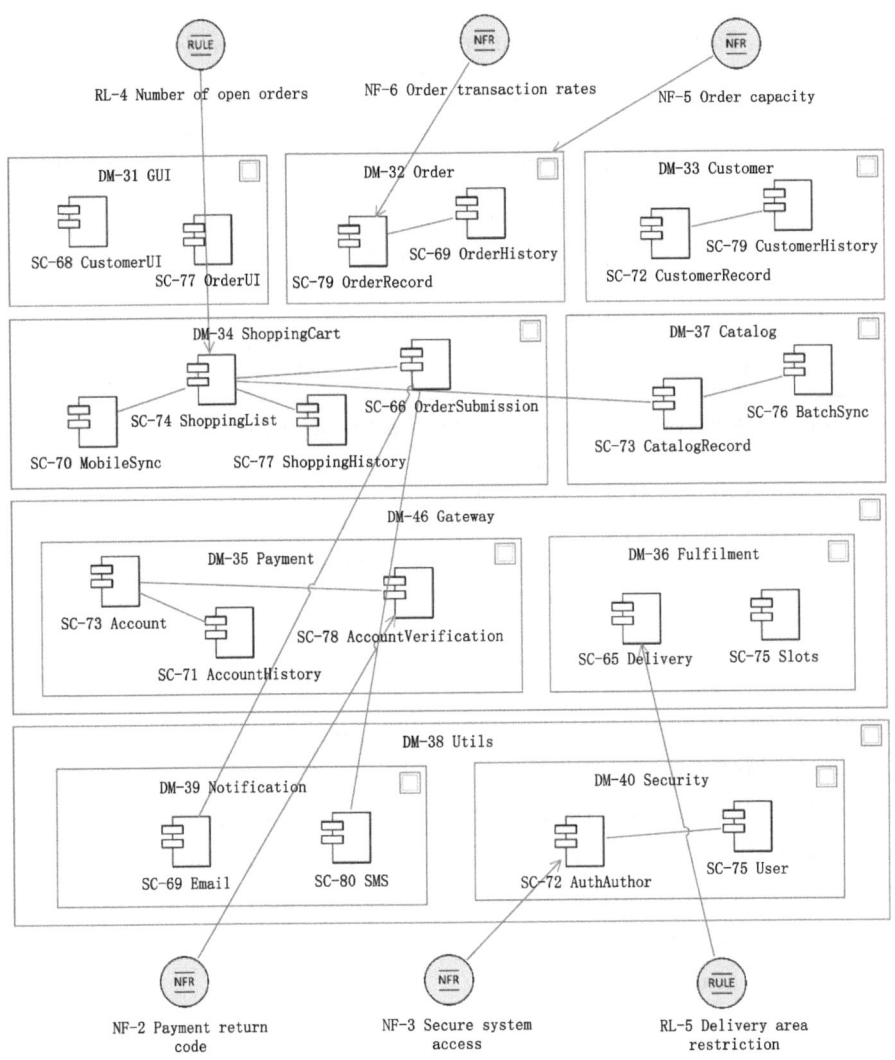

Figure 2-44. *Component association with the requirements*

Table 2-20. *Requirement Elements in Figure 2-43*

Element	Description
RL-4 Number of open orders	A customer can only have one open (unsubmitted) order at a time.
RL-5 Delivery area restriction	The delivery restriction is determined by the customer's zip code, which is the first two digits starting with 98 and 87.
NF-5 Order capacity	Orders are expected to increase by 56 million within a year.
NF-6 Order transaction rates	System is required to support 18000 orders per hour.
NF-3 Secure system access	Customer/user ID and passwords will be coordinated across channels.
NF-2 Payment return code	Payment method and order amount information will be submitted to the backend payment gateway system, which sends back a return code indicating success or failure (insufficient funds, expired card, etc.).

For the non-functional requirements in question, cross-checking and clarification must be done during the architectural process. For example, NF-5 does not specify the base number of orders, but it can be inferred or calculated from NF-6. NF-6 does not provide enough information to calculate throughput, which also requires information on data volumes (record size, etc.) and other data characteristics such as volatility, concurrency, and CRUD actions.

HINT

Significance of Requirements: Not all requirements are equal in the eyes of the solution architect. In general, most of the business functional requirements are not architectural concerns. Most of the business rule requirements and some of the non-functional requirements are not architecturally significant.

Data Domain Service Determination

Table 2-21 is *an initial data domain analysis* in a tabular form and
Figure 2-45 shows a data relationship view. Data tables are grouped and
partitioned into data domain, and database schemas based on business
logic and non-functional requirements. Architecturally, the partitioning
is primarily based on data characteristics, such as operational *order* data
and analytical *order history* data, as well as tradeoffs between scalable
distribution (BASE)[11] and transactional integrity (ACID).[12]

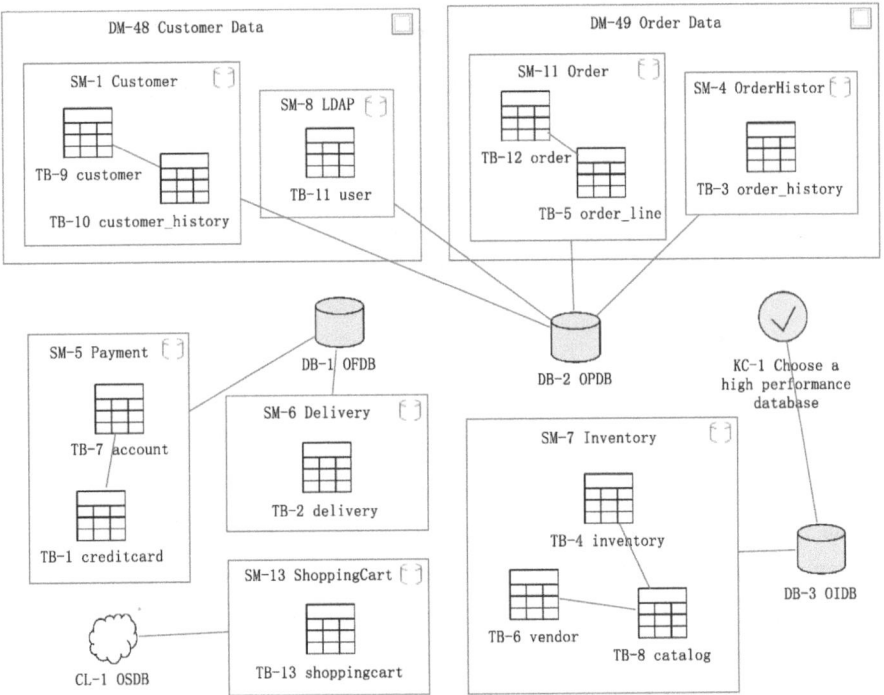

Figure 2-45. *SRV-9 static data relationship*

---

[11] Basically Available, Soft-state, Eventually consistent
[12] Atomicity, Consistency, Isolation, and Durability

Table 2-21. *Initial Data Domain Identification*

Database	Data Domain	Schema	Table
OPDB	Customer	Customer	customer
			customer_history
		LDAP	user
	Order	Order	order
			order_line
		OrderHistory	order_history
OFDB	Payment	Payment	creditcard
			account
	Delivery	Delivery	delivery
OIDB	Inventory	Inventory	vender
			inventory
			catalog
OSDB	CloudStorage	ShoppingCart	shoppingcart

As shown in Figure 2-45, each data domain has its own schema. However, based on a key architectural consideration (KC-1), the OIDB needs to have a high-performance database, which requires an iterative consideration.

Tables 2-22 and 2-23 show an architecture decision to choose the OIDB database in Figure 2-45.

Table 2-22. *Key Choice of NoSQL Database*

Element	Property	Description
KC-1 Choose a high-performance database	Assumption	Select a NoSQL key-value database for high-performance
	Alternative	1. Aerospike – SSD restriction 2. CouchBase – Better for document transaction 3. Apache Riak – Operationally simple but technically sophisticated 4. Redis – In-memory support with optional durability, lightweight, open-source and popular use
	Decision	Choice 4
	Rationale	See: NoSQL database comparative sheet (Table 2-23)
	Implication	Demerits: eventual consistency and data modeling capability of the chosen NoSQL database (less sought after in this solution)
	Status	Pending

Table 2-23. *NoSQL Key-Value DB Comparative Sheet*

Database	Data Model	Transaction Type	Clustering	APIs
Aerospike	Key-value	Strictly consistent single-server transactions and eventually consistent across a cluster	Sharding and replication	It has a subset of SQL language and drivers for most common languages
CouchBase	Key-value and JSON	Strict consistency for single-document transaction	Sharding and replication	REST-based API for Java and other languages
Riak	Key-value	Dynamo tunable consistency	Consistent hashing	REST API for Java, Ruby, etc.
Redis	Key-value	Strictly consistent within a single server	Master-slave replication	API for most used language

---

HINT

Multidimensional View of Data: The data relationship view reflects how each data service interacts and performs from business architecture, application architecture, and technical architecture. It is a cross-view area consideration. Each data service must be validated against its subject domain, data model, and business rules, data source, storage, and governance (standard, data integrity and confidentiality, etc.).

Component Service Realization

A-ESA can optionally provide realization guidance for critical service component implementations. Figure 2-46 exemplifies an implementation consideration for the ShoppingCart service.[13] This solution assumes a *key/value* store implementation that maximizes scalability, availability, and consistency.

---

[13] This service component realization utilized Azure Cloud implementation best practices.

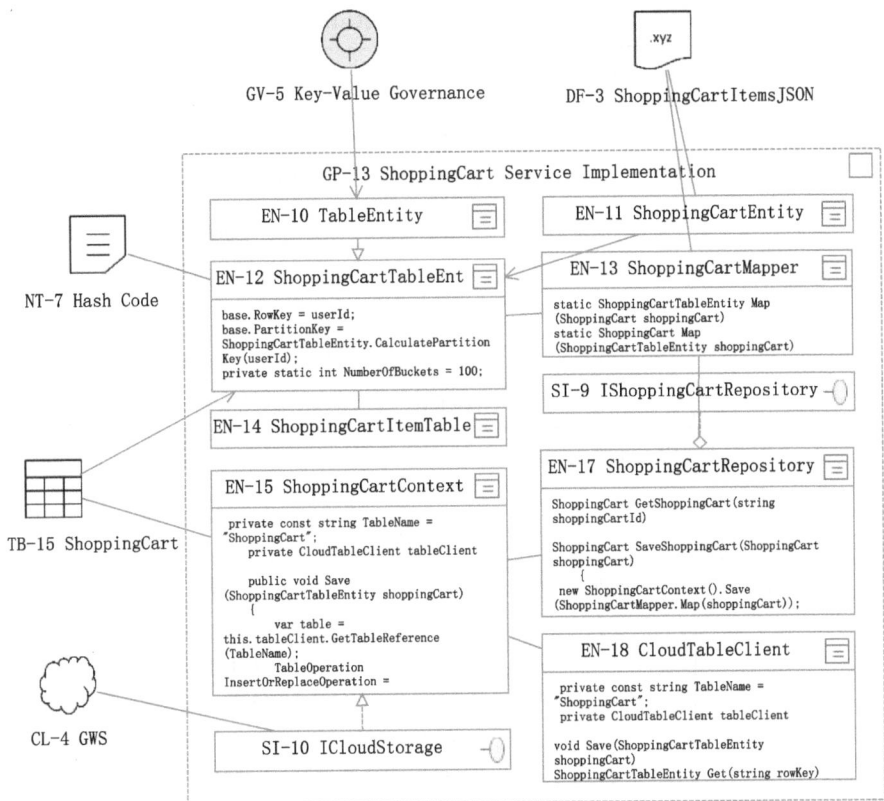

Figure 2-46. *ShoppingCart service implementation view*

Table 2-24 shows a *governance property specification* for the ShoppingCart service realization. Table 2-25 shows part of the code snippet specification for the ShoppingCart service realization.

Table 2-24. *Governance for the ShoppingCart Service Implementation*

Element	Property	Description
GV-5 Implementing a Key/Value Store for Scalability, Availability, and Consistency	Partition keys	Generate and populate Partition Keys in the `ShoppingCartTableEntity` class for scalability
	Data replication	Use cloud service to replicate the data in each storage account twice at its location to ensure availability
	Timestamp	Define that each item in a table has a Timestamp attribute to support optimistic concurrency
	Snapshot isolation	Use cloud service to implement snapshot isolation for non-blocking read operations to maximize consistency
NT-7 Hash Code	Hash code	The `CalculatePartitionKey` method in the `ShoppingCartTableEntity` class uses the `GetHashCode` method from `System.Object` to generate a hash value for the customer ID, and then uses modulus arithmetic to determine in which partition the shopping cart should reside

Table 2-25. *Code Snippet Property for the ShoppingCart Service*

Element	Description

EN-12 ShoppingCartTableEntity	```base.RowKey = userId;```
	```base.PartitionKey =```
	```ShoppingCartTableEntity.CalculatePartitionKey```
	```(userId);```
	```private static int NumberOfBuckets = 100;```
	```string ShoppingCartItemsJSON { get; set; }```
	```public Guid TrackingId { get; set; }```
	```public static string CalculatePartitionKey```
	```(string userId) {}```

(*continued*)

Table 2-25. (*continued*)

Element	Description

EN-15 ShoppingCartContext	`private const string TableName = "ShoppingCart";`
	`private CloudTableClient tableClient`
	`public void Save(ShoppingCartTableEntity`
	`shoppingCart) {} public void`
	`Delete(ShoppingCartTableEntity shoppingCart){}`
	`public ShoppingCartTableEntity Get(string`
	`rowKey) {}`
	`public void Delete(ShoppingCartTableEntity`
	`shoppingCart) {}`

EN-18 CloudTableClient	`private const string TableName = "ShoppingCart";`
	`private CloudTableClient tableClient`
	`void Save(ShoppingCartTableEntity shoppingCart)`
	`ShoppingCartTableEntity Get(string rowKey)`
	`void Delete(ShoppingCartTableEntity shoppingCart)`

EN-14
ShoppingCartItemTableEntity
	`public int Quantity { get; set; }`
	`public int ProductId { get; set; }`
	`public string ProductName { get; set; }`
	`public decimal ProductPrice { get; set; }`
	`public string CheckoutErrorMessage { get; set; }`

EN-17 ShoppingCartRepository	`ShoppingCart SaveShoppingCart(ShoppingCart`
	`shoppingCart)`
	`{`
	`new ShoppingCartContext().`
	`Save(ShoppingCartMapper.Map(shoppingCart));`
	`return shoppingCart;`
	`}`

Note:
– The code is for illustration purposes only and is intentionally truncated.

One of the A-ESA realistic tooling capabilities is to provide a link between architecture and design, as seen in Figure 2-46. However, such a capability should be used with caution: only for a limited number of significant cases or landing examples. Extensive use of design capability defeats the purpose of architecture, which usually remains at an appropriate level of abstraction.

Validation Views

Validation is a powerful way to cross-check element correlations, viewpoint relevance, relationship and direction filtering, orphaned or redundant elements/relationships, and depth analysis. The validation view can be presented as a functional view, an operational view, a cross-cutting view, or any view from a unique perspective.

Figure 2-47 shows an architecture overview assessment from the simplified AS-IS order management solution environment. As can be seen in Figure 2-47, there is no mobile access, no clearly defined interfaces such as payment service interface, no order process management, and no enterprise service hub. It is a typical monolithic solution.

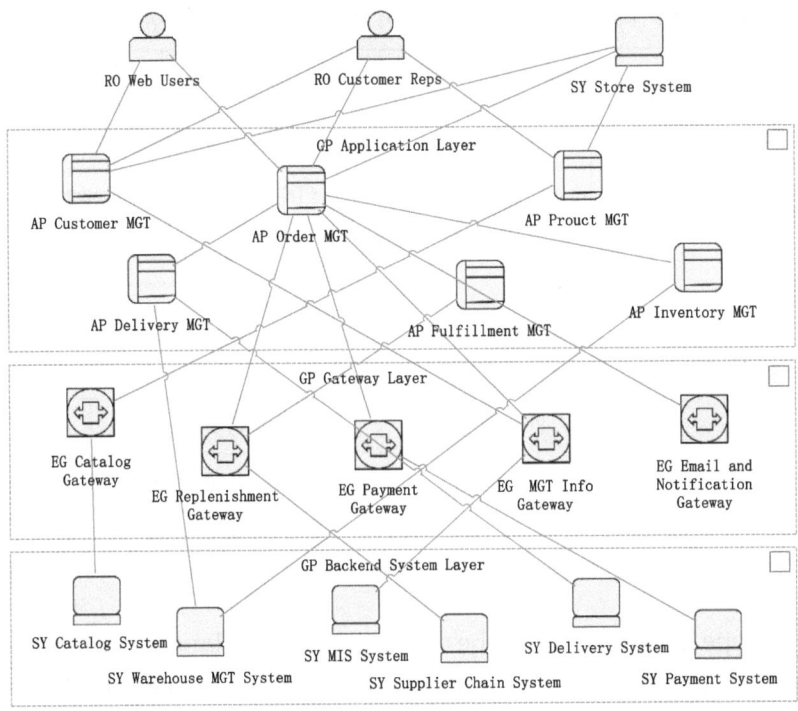

Figure 2-47. *AS-IS architecture overview validation*

Figure 2-48 and Table 2-26 show an initial validation view around the relatively straightforward *payment interface* element.

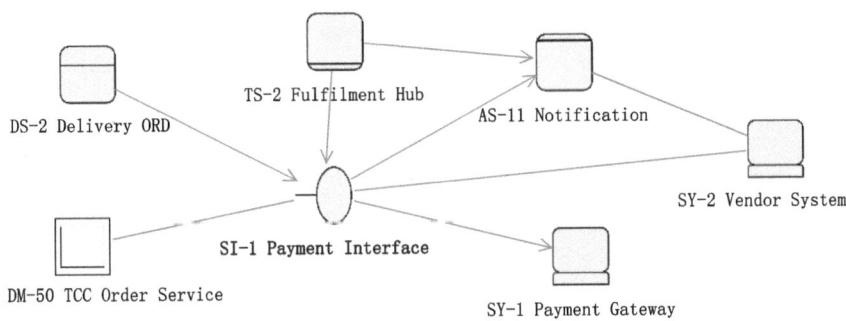

Figure 2-48. *SI-1 payment interface correlations*

Table 2-26. *Views Associated with the Payment Interface Element*

Element	Property	Description
SI Payment Interface	Associated View	AOV-2 Architecture Overview
		PTN-2 Distributed transaction
		SIV-1 Order Process
		SRV-7 Service Relationship
TS Fulfillment Hub	Associated Element	AS Notification
		SI Delivery Interface
		PS Oder Realization
		SI Payment Interface

This type of validation can be automated in the cloud environment or via vendor analytical tools, but it is still helpful for the cloud solution architects to have a better architecture view via ESA modeling when weighing tradeoffs such as cloud cost prior to deployment runtime. It is also helpful to track the *n-depth* correlation of an element.

A-ESA Profile 6: Microservice Architecture (MSA)

An MSA consists of service domain considerations, development environment, and cloud platform capability and operations, which together morph into the *three amigos* of microservice architecture: domain services, DevOps culture, and scalable virtual nodes.

This profile[14] demonstrates the two important parts of microservice architecture: *bounded context* identification and key considerations for the microservice *operational environment.*

Note that MSA is an iterative design process, better with event storming in an Agile approach. You can use your familiar iconic shapes or different colors to vividly represent MSA elements with related A-ESA notations, collaborating between business and IT teams.

Microservice Design Views

One of the challenges for microservice architecture is how to define the boundaries of the microservices. Microservices should be designed around the *business capabilities*, not horizontal layers such as data access or messaging. Therefore, it is recommended to use both intent-driven (*case scenario view*) and outcome-driven (*functional view*) approaches iteratively to achieve the desired result.

A microservice should follow the basic application service rule: right granularity, well-designed interface, loose coupling, and high cohesion. But putting this rule into practice requires deep architectural thinking in terms of business domain expertise, value goals, requirements, team size, and service level characteristics (SLCs).

---

[14] This profile referenced "Azure Architecture Center: Microservices architecture design" on top of the original OPS solution profile.

Initial Case Scenario View

Figure 2-49 shows an initial domain analysis of the solution application's functional requirements, including external system interfaces or third-party services associated with the order management system.

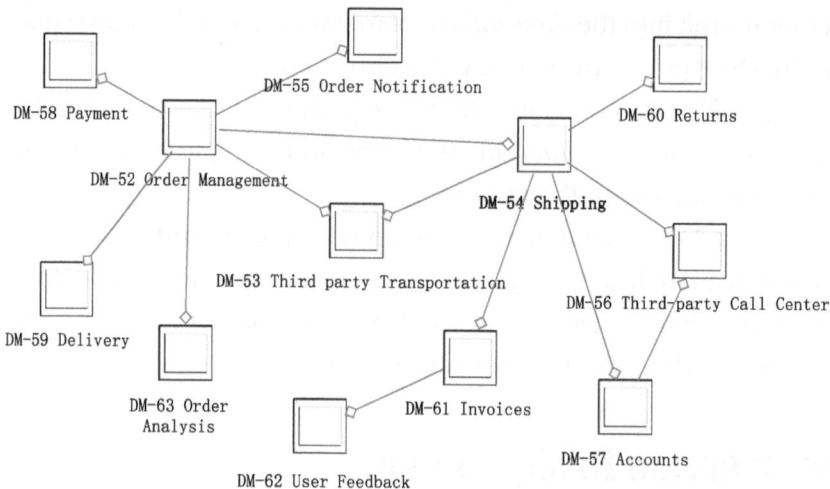

Figure 2-49. *An initial domain identification*

Figure 2-50 goes one step further to define the bounded contexts of domain. Each bounded context contains a domain model that represents a particular subdomain of the solution application.

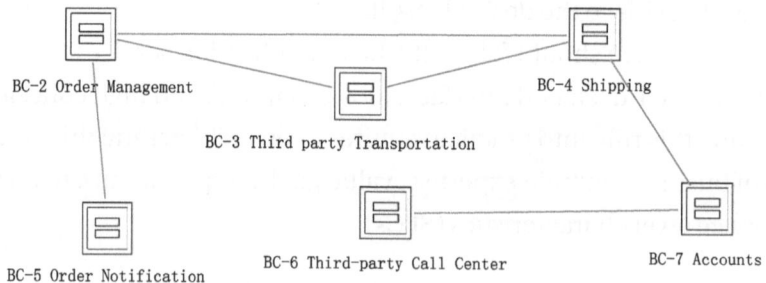

Figure 2-50. *Intended bounded contexts*

HINT

Bounded Context: The concept of a bounded context can be open to interpretation, as it is not always evident whether it pertains to a distributed service or simply a logical boundary within a monolithically deployed application. While defining a service for each bounded context is a good starting point, it is not always necessary to confine your design to this approach. In certain instances, it becomes imperative to devise a bounded context or business microservice that is constituted of a multitude of physical services. However, it is important to remember that the bounded context and the microservice are closely related.

DDD Pattern View

Figure 2-51 shows a high-level *DDD* (domain-driven design) pattern between domain services, aggregate, root entity, entity, and value object.

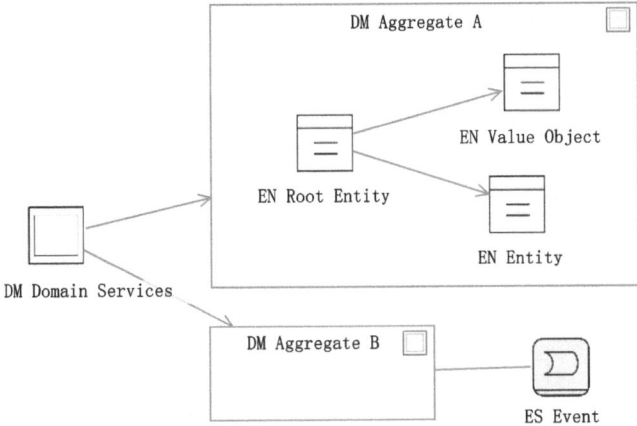

Figure 2-51. *Domain-driven design pattern*

Functional Service and Component Service Realization View

While the case scenarios (Figures 2-49 and 2-50) identify and build the domain model, the functional view over here moves it to the application design.

Microservice design, as the name implies, is an architectural design process, so the functional service view is created simultaneously and iteratively with the component service realization view.

This section defines aggregates (containing entities and value objects) and domain services in the *shipping* bounded context. Note that an aggregate is typically derived from business requirements, rather than technical concerns such as data access or messaging. An aggregate should have a high degree of functional cohesion and loose coupling.

Figure 2-52 shows the aggregates, entities, and value objects based on the DDD pattern (see Figure 2-51). The identified entities include *account, order, delivery, package, notification, confirmation,* and *tag info.* Among them, *account, order, delivery,* and *package* are aggregates, and *notification, confirmation,* and *tag info* are child entities. The identified value objects include *zip code, package weight,* and *package size.* Figure 2-53 shows the relationship between these entities.

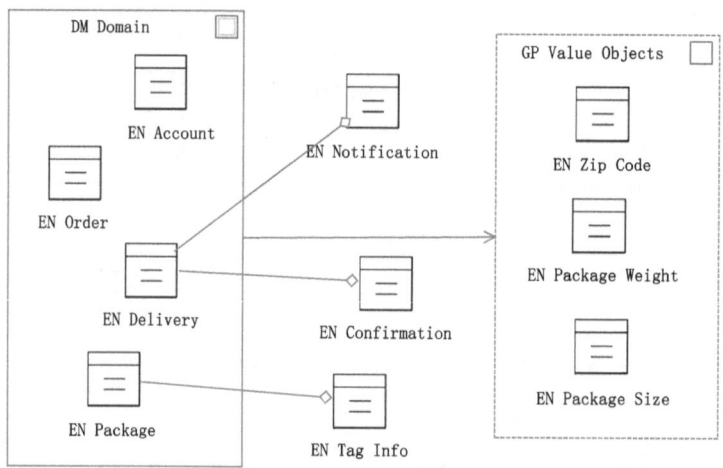

Figure 2-52. *Aggregate, entity, and value object identification*

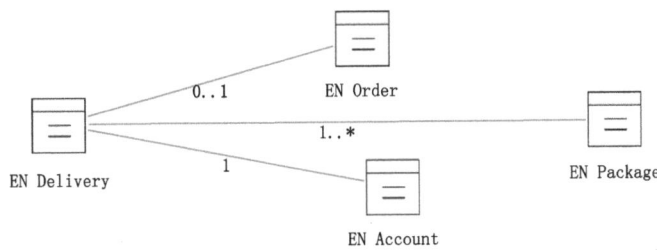

Figure 2-53. *Entity realization relationship*

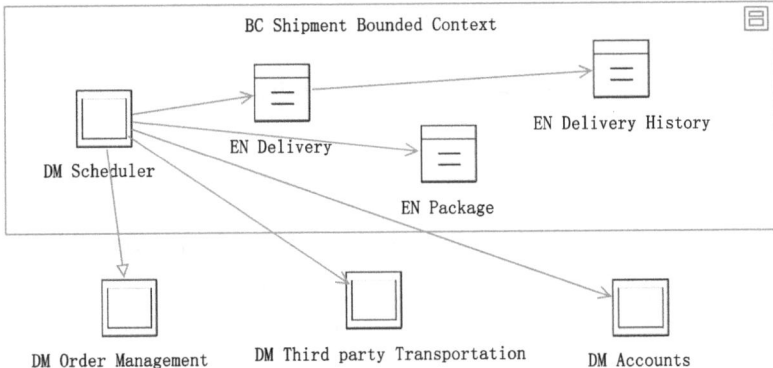

Figure 2-54. *Resulting shipping bounded context*

In general, the context of a microservice should be larger than an aggregate, and smaller than a bounded context. Figure 2-54 shows the resulting *shipping* bounded context.

As shown in Figure 2-55, the design and development team included domain services (*scheduler* and *supervisor*) and domain events (*order status* and *delivery status*). Table 2-27 shows the property descriptions of *delivery status*.

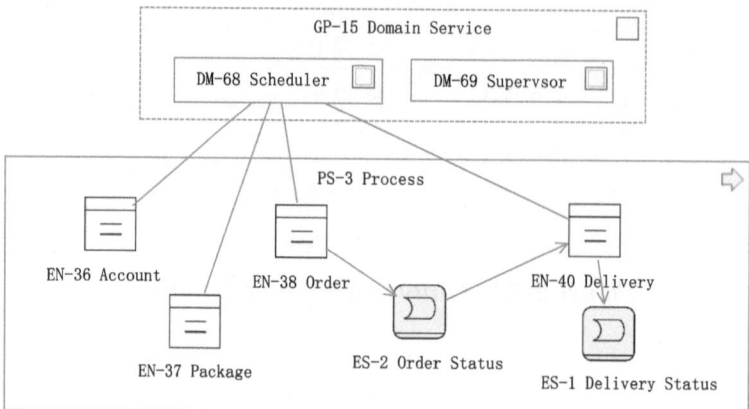

Figure 2-55. *Coordination process*

Table 2-27. *Property of Delivery Status Element*

Element	Property Name	Property Description
ES-1 Delivery Status	Status Indicator	Scheduled
	Status Indicator	Rescheduled
	Status Indicator	Cancelled
	Status Indicator	Completed

The bounded contexts identified in this step are further defined into microservices by factors such as NFR criteria, as shown in the following metrics view section.

HINT

Aggregate in DDD: An aggregate serves as a boundary for consistency and models transactional invariants. However, it is the responsibility of the application, not the data layer, to enforce the necessary invariants for the domain.

Value Object vs. Entity: Do not be confused by the terminology. In this context, "value object" refers to the concept of an entity in modeling.

Metrics Views

Figure 2-56 shows the metrics mapping for the microservice architecture.

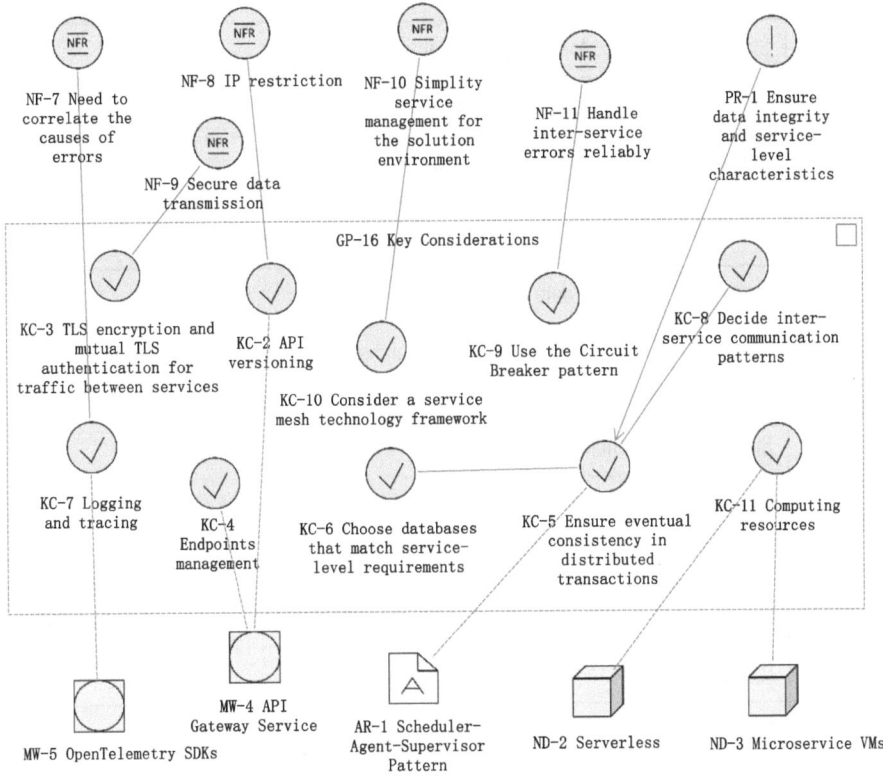

Figure 2-56. *Microservice key considerations*

For microservice architecture, scalability and high availability are easily addressed, while various other concerns such as *reliability* and *data integrity* become more challenging. Strong governance must be in place via critical metrics mapping and considerations (see Figure 2-56).

Table 2-28 displays the property specifications of the metrics elements depicted in Figure 2-56.

Table 2-28. *Metrics Element Specifications*

Element	Property	Property Specification
PR-1	Statement	Ensure data integrity and service-level characteristics
NF-7	Statement	Need to correlate the causes of errors
KC-7	Statement	Choose effective logging and traceability
	Decision	Use the standardized OpenTelemetry SDKs to enrich existing spans with entity keys (such as OrderID) tracing and enable context propagation (sending a correlation ID)
KC-3	Statement	Secure data transmission
	Decision	Implement TLS encryption and mutual TLS authentication for traffic between services
KC-9	Statement	Handle inter-service errors reliably
	Alternative	1. HTTP retry: The POST and PATCH methods are not safe because they are not guaranteed to be idempotent. 2. Circuit Breaker pattern: The fallback logic is employed to prevent a service from repeatedly attempting an operation that is likely to fail.
	Decision	Use the Circuit Breaker pattern
KC-8	Statement	Decide on inter-service communication patterns
	Alternative	1. Synchronous communication 2. asynchronous I/O (or asynchronous protocol)
	Decision	Internal services use synchronous call, and services that use asynchronous communication must be specified in the architecture view.

(*continued*)

Table 2-28. (*continued*)

Element	Property	Property Specification
KC-5	Statement	Ensure eventual consistency in distributed transactions
	Decision	Use the Scheduler-Agent-Supervisor pattern to coordinate *shipping* as a single operation using these supervisor attributes: – OrderID in the orders database. – LockedBy, the instance ID of the worker role handling the order. – CompleteBy, the time by which the order should be processed. – ProcessState, the current state of the task handling the order. The possible states are pending, processing, processed, error, and failure_count.
KC-10	Statement	Consider a service mesh technology framework
	Rationale	A service mesh incorporates intelligent load balancing algorithms based on observed latency or other performance metrics. It handles service-to-service communication, layer-7 routing, retry of failed requests, circuit breaking, metrics of inter-service calls, and mutual TLS authentication for service-to-service calls. However, it has these drawbacks: – It currently works for container orchestrators, rather than serverless. – Adding this feature will make cluster setup and configuration more complex, which could negatively affect performance.
	Decision	Don't use service mesh for this architecture.
KC-11	Statement	Maintain self-control over computing resources
	Alternative	1. The orchestrator for VMs provides greater control over service configuration and cluster management, but this comes at the cost of increased complexity. 2. Serverless computing requires no management of VMs or virtual network infrastructure, and scales automatically. However, this approach is better suited for small, granular functions coordinated by event-based triggers and may not be appropriate for complex applications.
	Decision	Use both approaches. For functions that are relatively standalone, use serverless.

(*continued*)

Table 2-28. (*continued*)

Element	Property	Property Specification
KC-2	Statement	Ensure smooth API versioning
	Alternative	1. Expose both versions of the same service. 2. Run two versions of the service side by side, and route requests to one or the other version.
	Decision	Choose Option 2 and deprecate the old versions promptly. In addition, make sure that the API owner works with other teams to migrate to the new version.
KC-4	Statement	Choose an API endpoint management technology
	Decision	Use a mature API gateway that supports basic requirements such as Layer 7 routing, IP restrictions, and authentication, as well as the following API styles: – GraphQL – Restful API – Streaming API – JavaScript API – SDK – Asynchronous messaging – RPC API – Soap API – Hypermedia
KC-6	Statement	Choose databases that meet the required service level.
	Decision	Each microservice has its own data store (Table 2-31).

Note:
– In a microservice architecture, inter-service communication presents a challenge in balancing consistency and coordination. Think carefully about this important decision.

Not all architectural considerations need to be recorded, as they are likely already handled in the solution middleware or platform and are routinely implemented. Table 2-29 shows the architecture decisions from a *viability* perspective. Table 2-30 shows the architecture decisions by non-functional requirement category.

Table 2-29. *Motivation of Architecture Decisions*

KC	Statement	Business Impact	Arch Impact
KC-7	Choose effective logging and tracing	High	High
KC-3	Secure data transmission	High	
KC-9	Handle inter-service errors reliably		High
KC-8	Decide inter-service communication patterns		High
KC-5	Ensure eventual consistency in distributed transactions	High	High
KC-10	Consider a service mesh technology framework		High
KC-11	Maintain self-control of computing resources		High
KC-2	Ensure smooth API versioning	High	High
KC-4	Choose an API endpoints management technology		High

Table 2-30. *Architecture Decision by NFRs*

Element	Statement	M	S	R	P
KC-7	Choose effective logging and tracing	X			X
KC-3	Secure data transmission		X	X	
KC-9	Handle inter-service errors reliably			X	
KC-8	Decide Inter-service communication			X	X
KC-5	Ensure eventual consistency in distributed transactions			X	
KC-10	Consider a service mesh technology	X			X
KC-11	Self-control computing resources	X			
KC-2	Ensure smooth API versioning			X	
KC-4	Choose an API MGT technology	X	X		X

Note:
— M: Manageability, S: Security, R: Reliability, P: Performance

Data Collaboration Views

The main challenge of a microservices architecture is data collaboration. It is much harder than traditional monolithic architecture in two areas: 1) data query from multiple microservices while avoiding chatty communication, and 2) data consistency across microservices. For chatty APIs, service boundaries must be redrawn, and for transaction integrity, apply patterns such as *scheduler agent supervisor* and *compensating transaction* must be applied to keep data consistent across multiple services.

Supervisor Pattern View

Figure 2-57 shows the *scheduler-agent-supervisor* (or supervisor) pattern that is commonly referenced in the microservice cloud architecture.

This pattern enables retry actions caused by transient or long-lasting failures in a distributed enterprise application. If the failure is not easily recovered, the application is returned to a consistent state to ensure integrity.

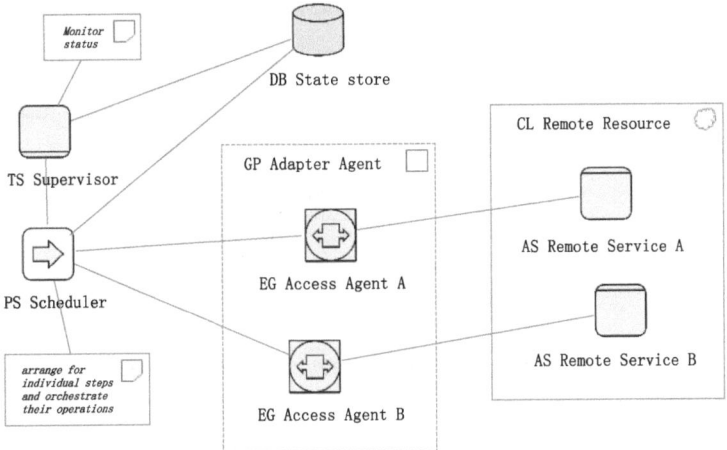

Figure 2-57. Scheduler-agent-supervisor pattern

Initial Data Service View

This step forms microservices from the previously identified bounded contexts. At this point, microservice decisions must be made based on implementation and operational impact. Note that service boundaries aren't fixed. As an application evolves, a microservice may be split into several smaller services.

As shown in Figure 2-58, this microservice view highlighted transactional integrity concerns using the *supervisor* pattern (see Figure 2-57) and the database selections in Table 2-31.

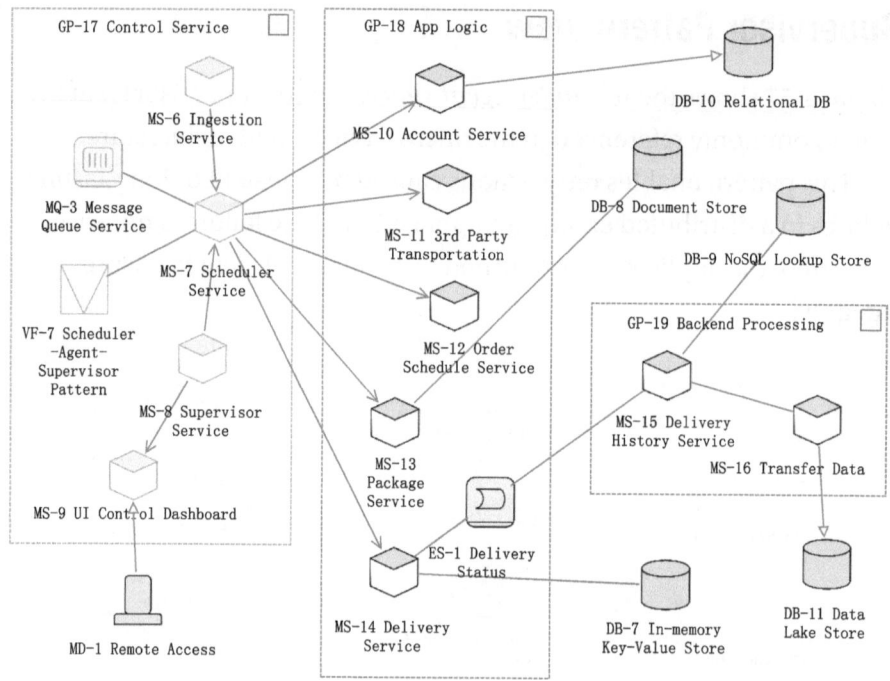

Figure 2-58. *Data service relationship*

Table 2-31. *Database Considerations*

Microservice	Usage	SLR	CRUD	Data Store
MS-13 Package Service	Long-term document storage	High volume and high throughput	CRUD	Sharded Document Store
MS-10 Account Service	Relational data storage	Reliability	CRUD	Relational DB
MS-14 Delivery Service	Simple query, in-flight operations	High throughput	CRUD	In-memory Key-Value Store
MS-16 Transfer Data	Data storage	Capacity	C	Data Lake Store
MS-15 Delivery History	Data lookup query and analysis	Fast read	R	NoSQL Lookup Store

In short, the key decision at the service boundary is transactional integrity and data consistency. For strong consistency, the required functionality must be contained in a single microservice. When in doubt, start with *coarse-grained* microservices.

HINT

Data Service Representation: Traditional data modeling uses the "one fact in one place" rule, whereas in a microservices architecture, a service might represent the source of truth for a given entity exposed through an API.

Polyglot Persistence: Each microservice requires the use of its own database schema. This approach naturally leads to polyglot persistence.[15]

Operational Environment Views

Designing communication across microservice boundaries is another challenge for microservice architectures from an application operations perspective.

In this step, non-functional requirements (including scalability, availability, performance, and security) from the DevOps environment are further considered. Through walkthrough analysis or operational simulation, ensure that there are no inter-dependencies and chatty calls between services at runtime, and that there is no network overhead from calling directly into the other bounded context. Also, consider the likely team structure issues. These issues may lead you to create a *mediator* service between the two contexts, or to redesign the microservices.

---

[15] Multiple data storage technologies in a single application

As shown in Figure 2-59, inter-service communications (such as synchronous vs. asynchronous) are expressed as connection relationships, and throughput considerations are implemented in the *ingestion* service which places incoming requests in a queue buffer. Leveraging service-oriented architecture (SOA), the service bus serves as the connection hub and the workflow includes microservice orchestrations.

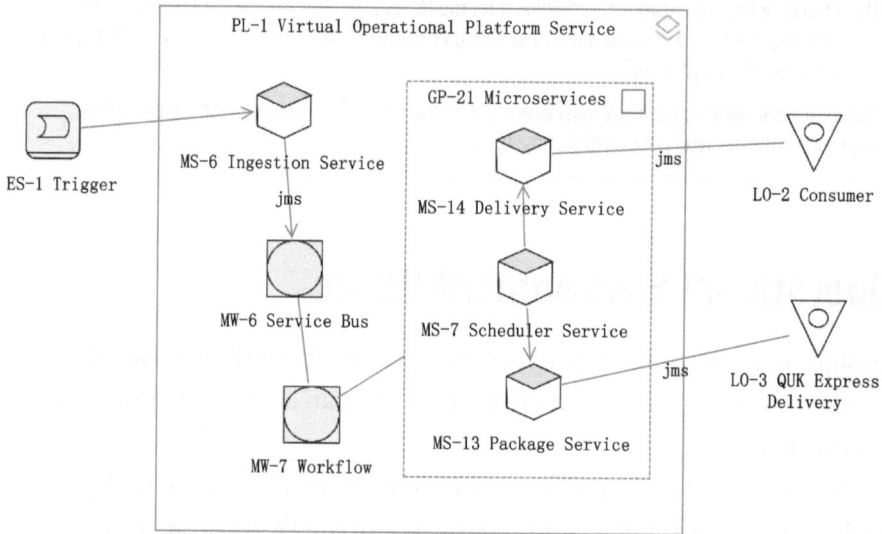

Figure 2-59. *Operational flow view*

Some of the key challenges of service-to-service communication need to be further addressed, such as resiliency (retry or circuit break) and distributed tracing as considered in the metrics view. In addition, for container applications, state and data management need to be further elaborated in the operational deployment views.

A-ESA Profile 7: Cloud Hosted Architecture (FaaS)

This solution profile presents a *cloud-hosted architecture* (CHA) *Function as a Service* (FaaS) MVAM[16] solution, to illustrate how a cloud-based event-driven architecture, in this case a very small solution, can leverage A-ESA modeling for a simpler yet clearer solution architecture specification.

Although the FaaS services are relatively more granular than microservices, the entire architecture associated with these services is worth exploring for cost effectiveness, tradeoff analysis, scalability, or cloud migration justification.

Note that each of the FaaS features can be modeled as a specific focus, or a specific NFR concern such as data management strategy, security impact, and resilience mechanism can be individually profiled or as a special set of views.

This solution presentation has used some AWS[17] images (see Table 2-32). In cloud-based architecture, there are many cloud middleware. To distinguish them for better architectural understanding, a special set of image notations is recommended instead of the middleware notation from the default A-ESA palette.

---

[16] Minimal Viable Architecture Model

[17] Amazon Web Service

Table 2-32. *A List of Special Cloud Middleware Images*

Element	Image	Description
MW API Gateway		A fully managed service that makes it easy for developers to create, publish, maintain, monitor, and secure APIs at any scale.
MW CloudWatch		A monitoring service designed for DevOps engineers, SRE guys, IT managers, and product owners.
MW Lambda		A serverless platform that offers the convenience of running your code on a high-availability compute infrastructure.
MW S3		Cloud object storage with industry-leading scalability, data availability, security, and performance.
MW SNS		A managed service that provides message delivery from publishers to subscribers.
MW Lambda Function		Takes one or more functions as arguments or returns one or more functions.
MW Rekognition		A cloud-based image and video analysis service that lets you add advanced computer vision to your apps.

HINT

FaaS vs. Serverless: FaaS is a form of serverless computing. As part of the cloud architecture, the serverless architecture has a broad scope, including UI, publish/subscribe infrastructure, workflow/orchestration, active databases, API gateways, system management, and data services. Taken together, these technologies can be overwhelming. A-ESA is also well suited to this type of architecture.

Capability Views

Capabilities can have a *business focus* or an *IT focus*. Figure 2-60 illustrates some of the cloud service IT capabilities used in the FaaS solution.

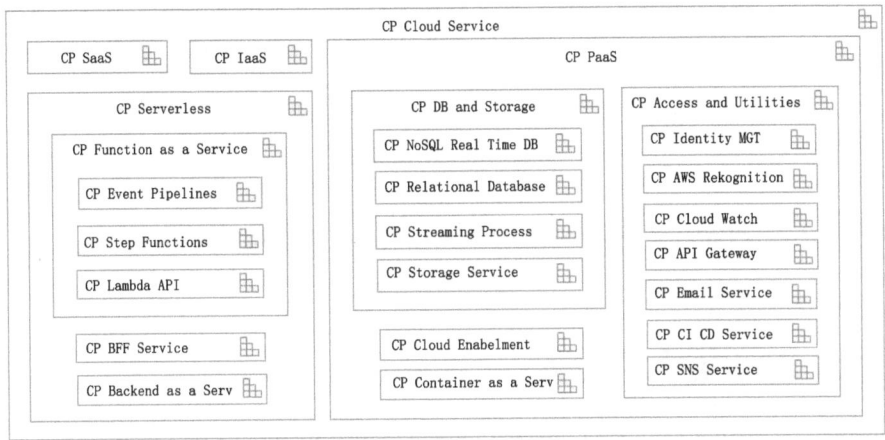

Figure 2-60. *Part of cloud capabilities*

Capabilities can be associated with generic IT services as an offering. The cloud platform offers not only Function as a Service (FaaS), but also *Construct as Code* (CaS).[18] In A-ESA, this type of service that is not in exact component or interface form is called a virtual service. Of course, the category depends on your solution definition; for example, the event-related constructs may belong to the event service. Figure 2-61 shows an example of the cloud constructs capabilities.

---

[18] Construct as Code means either Composition as Code or Configuration as Code

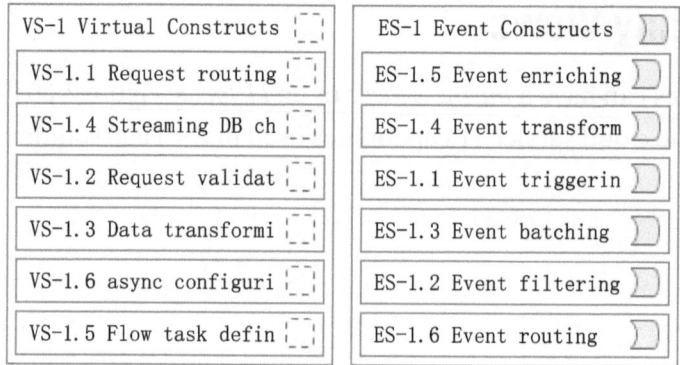

Figure 2-61. *Constructs virtual service capabilities*

Case Scenario View

Figure 2-62 and Table 2-33 show an application process view of the FaaS solution. This is a simple image processing solution. This case scenario is more from a system use case perspective. Together with DevOps and continuous validation, it works as both a requirements view and a system solution view. In other words, fewer and simpler architecture views are recommended for a FaaS application.

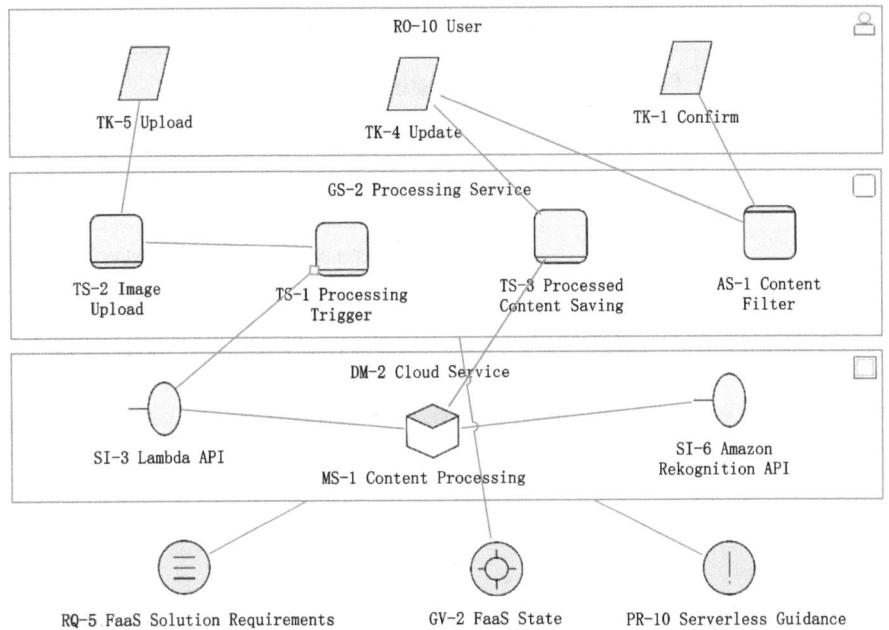

Figure 2-62. *A FaaS process view*

Table 2-33. *Partial Properties of FaaS Application Process*

Element	Property	Property Specification
RQ-5 FaaS Solution Requirements	Requirement 1	Utilize AI image recognition features
	Requirement 2	Allow large volume processing
	Requirement 3	Enable multiple format support
GV-2 FaaS State	Guideline 1	Write stateless FaaS functions

(continued)

Table 2-33. (*continued*)

Element	Property	Property Specification
PR-10 Serverless Guidance	Principle 1	Enforce data consistency
	Principle 2	Separate application logic from technology
	Principle 3	Use a simpler model to automate infrastructure

Note
– Metrics elements can be mapped together in a metrics view or correlated in other views, starting with the case scenario view.
– On a large scale, events can become a bottleneck for the system, so consider using aggregation or batching to reduce the overall load.

HINT

Principle vs. Requirement: The applicable strategic motivations from EA need to be specialized into concrete metric elements, including requirements, which can be derived from principles, while principles can be derived as generalizations of requirements.

Metrics View

Figure 2-63 is the governance view for the FaaS solution. The view manifests major metrics (governance, requirements, and constraints) to make decisions for architectural conformance. Table 2-34 contains some properties of the governance view.

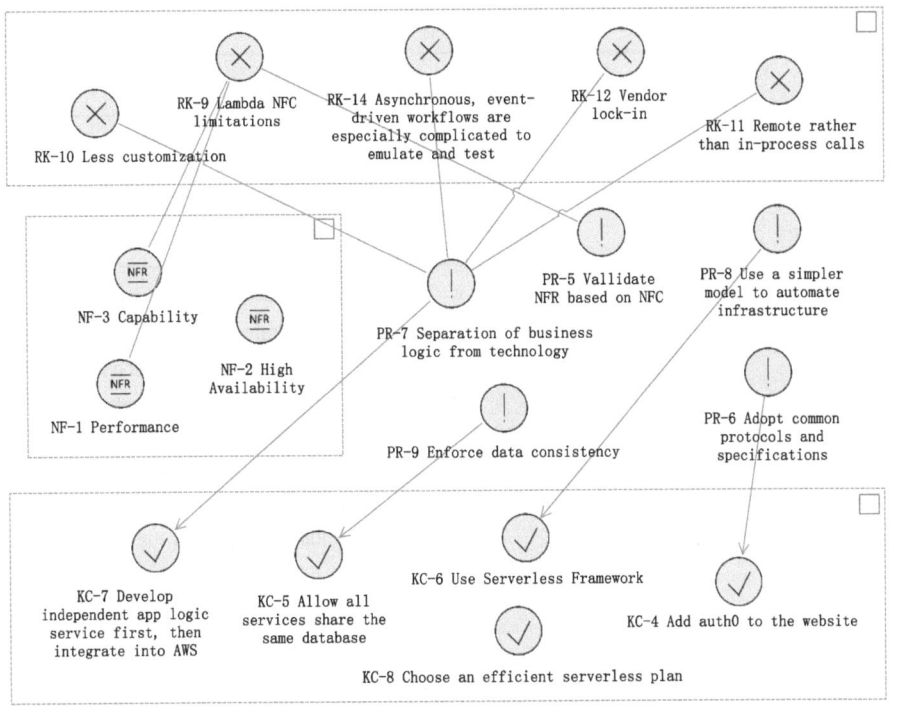

Figure 2-63. *A FaaS governance view*

Table 2-34. *Example Properties of the Governance View*

Element	Property	Property Specification
NF-3 Capability	Ephemeral disk capacity (/tmp space)	512MB
	Number of file descriptors	1024
NF-1 Performance	Number of processes and threads	2048
	Maximum execution duration	300 seconds
	Invoking request body payload size	6MB
	Invoking response body payload size	128K

(continued)

Table 2-34. (*continued*)

Element	Property	Property Specification
NF-2 High Availability	Availability	99.5%
KC-6 Use Serverless Framework	Key Choice	Choose a serverless development platform
	Alternatives	1. CloudFormation 2. Terraform 3. Serverless Application Model 4. Serverless Framework 5. AWS Application Composer
	Reason of Choice	Serverless Framework contains the following required runtimes: – aws-nodejs – aws-python – aws-java-gradle – aws-java-maven – aws-scala-sbt It provides FaaS functions, events and resources efficiently and safely. The other choices are either not optimized or less flexible.
	Decision	Serverless Framework
KC-8 Cost-saving on serverless platform	Statement	Choose an efficient serverless plan
	Decision	Leverage serverless cold start to save costs

Enterprise DevOps View

The DevOps view is generally a must in an Agile native-cloud architecture where development and deployment are done in a CI/CD[19] fashion, following the Agile principle of "you build, you run." This view reflects the connections between enterprise organizational capabilities, development portfolios, and deployment automation.

Figure 2-64 is a DevOps view for the FaaS solution development, testing, and deployment stages. Tables 2-35 and 2-36 show some properties related to the DevOps platform view.

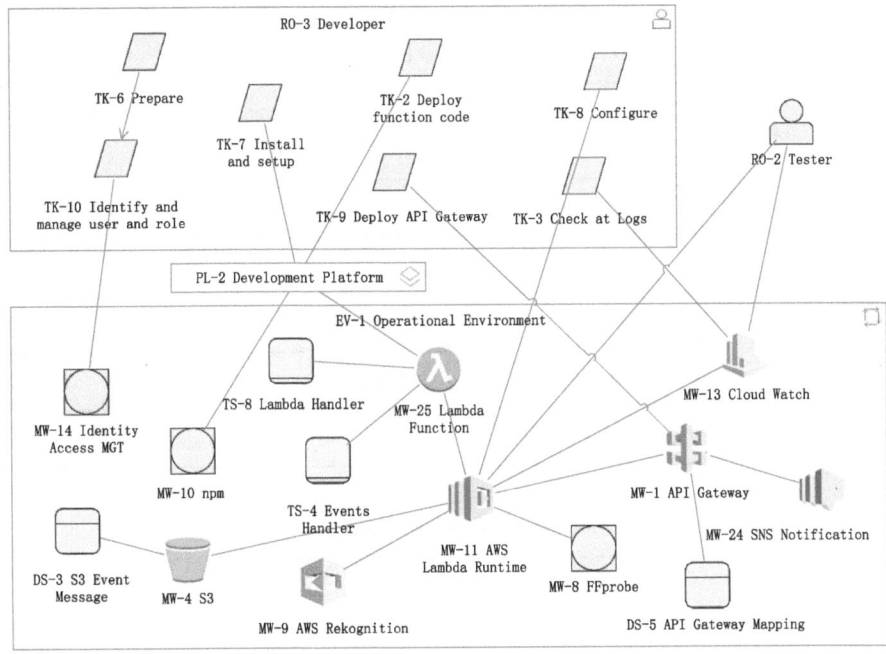

Figure 2-64. *DEV-4 DevOps view*

---

[19] Continuous integration/continuous deployment

Table 2-35. *DevOps View Properties*

View	Property	Selection Description
DEV-4 Enterprise DevOps View	Deployment Strategy	– In-place deployment – Rolling deployment – Blue-green deployment – Red-black deployment – Immutable deployment
	Testing Stage	– Prototype – Unit Test – Staging – QA Test – UAT
	Test Coverage	The target for automated integration testing is over 60%-line coverage.
	Project tracking	Jira
	Source control	Git
	Build and deployment	Maven/Jenkins
	Code quality	SonarQube
	Configuration MGT	Ansible

Note:
– The DevOps tool chain can also be expressed in the view if necessary.

Table 2-36. *Selected Element Properties of the DevOps*

Element	Property Name	Property Specification

MW-25 Lambda Function Deployment	Deployment Script	`"scripts": {`
		`"dp_cts": "run-local-lambda --file start.`
		`js --dp_cts-event/cts.json",`
		`"deploy": "aws lambda update-function-`
		`code --function-name --zip-file file://`
		`DPCustomTransSvc.zip",`
		`"predeploy": "zip -r DPCustomTransSvc.zip`
		`* -x *.zip *.json *.log"`
		`}`

DS-5 API Gateway Mapping	Map Request	Map API Gateway API Request Payload
	Map Response	Map API Gateway API Response Payload
	Map Format	Follow API Gateway JSON Schema format
DS-3 S3 Event Message	Event-Time	2020-02-02-T00:00:00.000Z
	Request-Parameters	SourceIP: 123.5.0.1
	Sequencer	UTUI86IYUY787588889YT
PL-2 Development Platform	Package-Manager	Npm: Manager Lambda functions and its dependencies
	Deployment	AWS Command Line Interface (CLI)
	Automation	Automated process for testing, packaging, and deploying Lambda functions to AWS
TS-8 Lambda Handler	Runtime	Node.js 4.1
	Handler	TransContent.handler
	Role	Execution-role
TS-4 Events Handler	Events	Object Created (All)
	Send To	Lambda Function

(*continued*)

Table 2-36. (*continued*)

Element	Property Name	Property Specification
MW-11 Lambda	Pipeline-ID	48878886993-yueuye
	Status	Active
	Input-Bucket	upload-translation-content
	ARN	arn:aws:elastictranscoder.us-mid-east-2-1 :8676345:pipeline/48878886993-yueuye
MW-25 Lambda Function	Handler Syntax	exports.handler = function(event, context, callback)
	Event object	JSON object that contains information about the event and the source that triggers the Lambda function.
	Context object	It contains useful information about the Lambda's runtime
	Callback function	An optional parameter that returns information to the caller
MW-4 S3 Upload Bucket	File-Name	uploaded_image_83687.png
	Preset-ID	3454543555-56565655

HINT

DevOps Maturity Level: The adoption of FaaS facilitates the DevOps maturity level (basic, emerging, coordinated, enhanced, and dynamic).

Deployment Packages in the Cloud: In a public cloud environment, deployment becomes easy because it is not required to take care of certain tasks of middleware installation, configuration and operational design. But a clear deployment unit package mapping still helps manage non-functional requirements and future scaling.

Functional Collaboration View

Figure 2-65 shows a file processing collaboration view over the serverless platform. Since much of the functional work can be automated, the focus is now on the process efficiency. Of course, without a clear architectural picture, you are likely to build an application in the dark that will not scale well or operate without cost optimization.

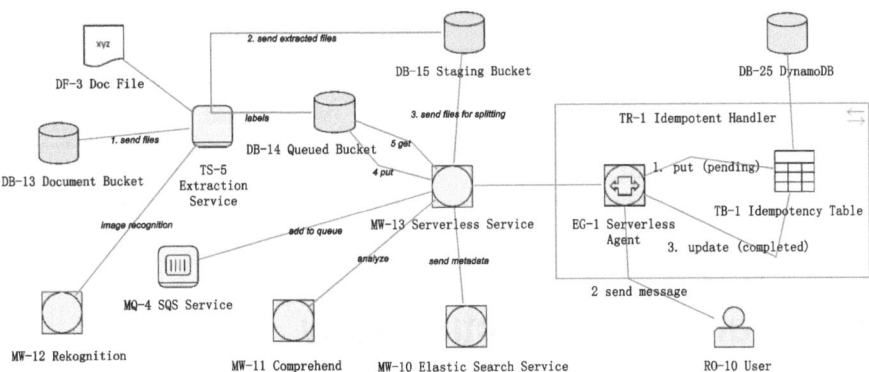

Figure 2-65. *Serverless file processing collaboration view*

As seen in Figure 2-65, the similar *idempotency* function implemented in Profile 5 is now performed by the serverless platform through automation or tool configuration. Table 2-37 shows the illustrative configuration properties.

Table 2-37. *Idempotent Configuration*

Element	Property	Description
TB-1 Idempotency Table	Type	DynamoDB::Table
	AttributeDefinitions	– AttributeName: id – AttributeType: S
	KeySchema	– AttributeName: id – KeyType: HASH
	TimeToLiveSpecification	– AttributeName: expiration – Enabled: true
	BillingMode	PAY_PER_REQUEST

Operational Infrastructure View

For the cloud platform, the operational infrastructure is taken care of by the PaaS[20] and IaaS, so the focus can be on the application development and deployment side. However, you still need to know the operational architecture for clear understanding and cost optimization. Because cloud architecture uses many architectural patterns or visualization capabilities, it is easy to see what is behind the scenes. Figure 2-66 shows an operational infrastructure model view in a serverless platform environment.

---

[20] PaaS can be further divided into aPaaS (more at an application level) and iPaaS (more at an infrastructure level).

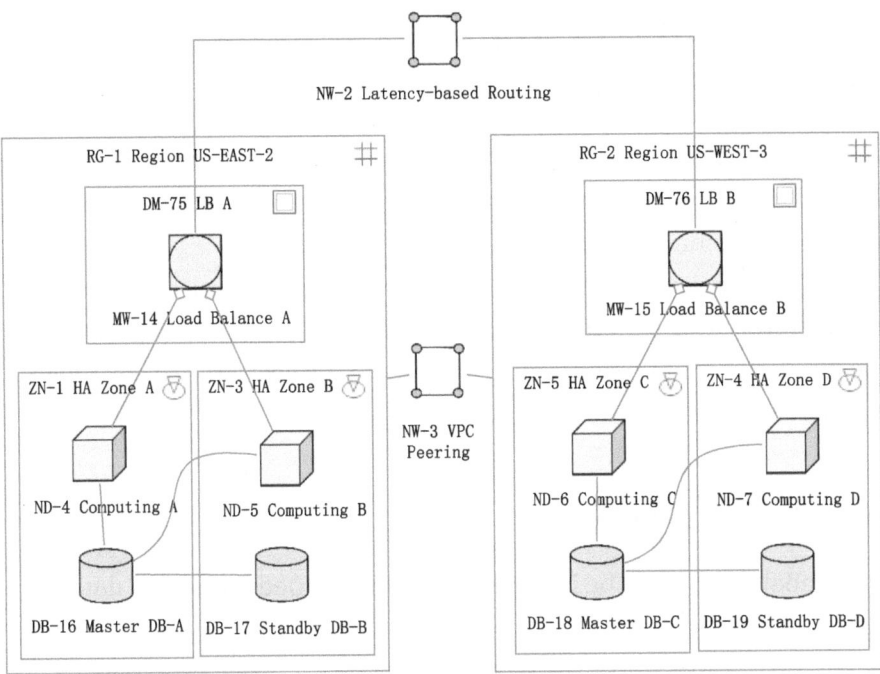

Figure 2-66. *Serverless HA platform environment*

HINT

Cloud Design Pattern: In the cloud platform, you can leverage cloud design patterns when considering operational characteristics. For example, you can use the storage index pattern and write proxy pattern for data upload, the floating IP pattern and multi-server pattern for HA, the sharding write pattern and in-memory cache pattern for databases, and the on-demand NAT pattern and multiple load balance pattern for networking.

A-ESA Profile 8: Big Data Technical Architecture

Big data analytics and artificial intelligence usually fall under data architecture. However, big data architecture is a cross-cutting concern more on the technical side.

This enterprise solution architecture profile serves as a technical architecture, along with related data architecture and AI considerations. The profile covers functional and operational infrastructure from a *high-level overview* perspective.

Capability Views

Basic capabilities must be identified and determined for the big data environment.

Figure 2-67. *General Big Data architecture capabilities*

Big data architecture is a framework that defines a set of components, middleware products, and processes. Figure 2-67 shows a list of common big data architecture model with AI capabilities.

Case Scenario Views

For big data architecture, case scenarios are commonly presented as system use cases, data collection and streaming scenarios, and various data flow requirements. Figure 2-68 shows a *big data flow* process view. The client sends a request to a control node that assigns data node information for the process file, and the control node divides the file into multiple blocks and writes them to data nodes.

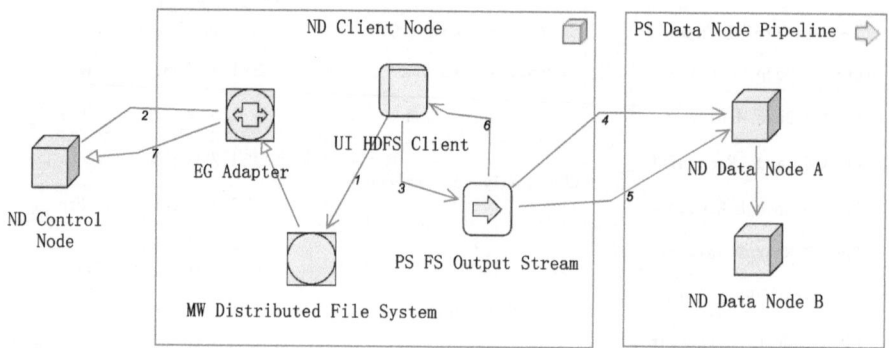

Figure 2-68. *Sample data flow process view*

Solution Architecture Overview

As big data architecture has a heavy use of middleware integration. Solution architecture overview is a focal view to see its constituents and collaborations.

Architecture Block View

Figure 2-69 shows a typical big data process architecture block view using a Hadoop ecosystem.

Figure 2-69. *Hadoop solution architecture block view*

The solution architecture overview can be further elaborated using
a relationship view, which can specify relationships between different
middleware such as the relationship between HBase and HDFS, between
HBase and Zookeeper/MapReduce, between HBase and Pig/Hive,
between HBase and Squoop/RDBMS, and between HBase and Hadoop.
Depending on the solution intent, either an application integration
architecture or a data architecture can be planned.

Architecture Layer View

The big data solution architecture generally covers four areas (see Figure 2-70): *data acquisition, data stream processing, data batch processing,* and *data analytics.* Or it can be divided into *data acquisition, data pipeline processing, data storage,* and *data analytics.* Each of these can be elaborated by its drilldown views. The architecture is determined by its own set of metrics based on service quality requirements (scalability, performance, security, and flexibility), governance, unique criteria (*volume, velocity, variety,* and *veracity*), and associated organizational structures and values.

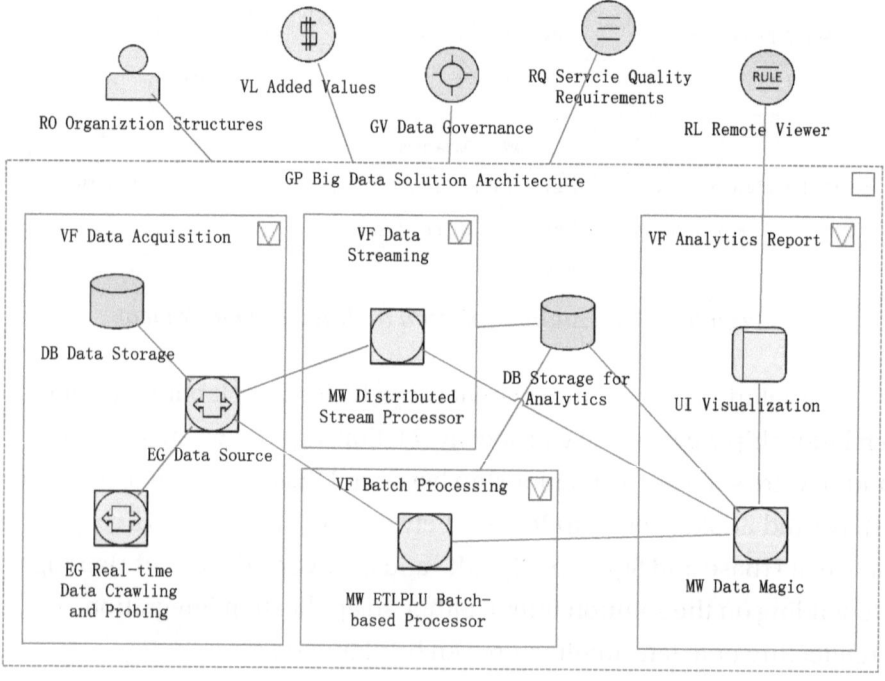

Figure 2-70. *Big data solution architecture overview*

From a solution architecture perspective, data integration and cross-cutting concerns are the key considerations. Figure 2-71 is an overview of an order data management integration associated with the big data architecture.

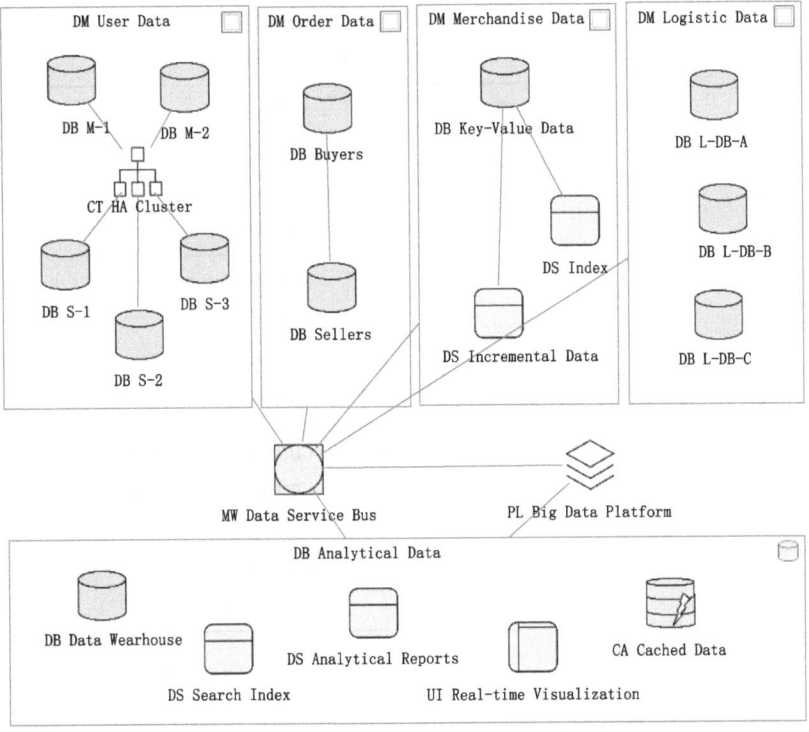

Figure 2-71. *AOV-7 order data management view*

Table 2-38 shows the property descriptions considered for the *order data management view* (see Figure 2-71).

Table 2-38. *Properties of Order Data Management*

View	Property	Description
AOV-6 Order Data Management View	DB Storage Option	– Apache HBase – Azure Tables – DynamoDB – MongoDB
	Storage Type	– NoSQL – Data warehouse – Cache – File-based
	DB Data Usage	– User session data – Transaction data – Business flow data – Dashboard data
	Data HA Option	– CDNs for static content scaling – Caching at different layers – NoSQL stores – Sharding – Read replicas

Architecture Platform View

Figure 2-72 shows a customer environment connected to a big data processing cloud platform such as *confluent cloud*.

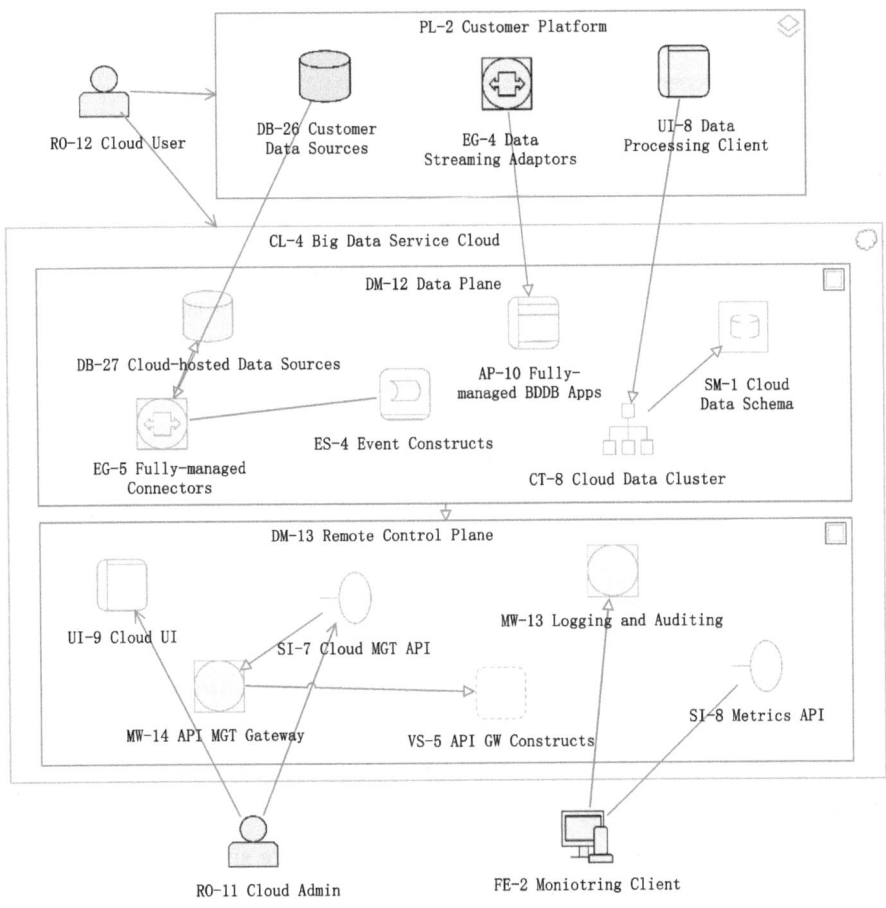

Figure 2-72. *Big data hybrid cloud*

The big data cloud platform is a specialized way to handle data processing needs. By leveraging the cloud capabilities, there will be no need to architect the many functional and operational services, but there still need to have a clear picture of what and how to connect, middleware choices, configurations, and so on. Table 2-39 shows an example *Construct as Code* (CaS) for some associated elements.

Table 2-39. *Cloud Construct Properties*

Element	Property	Description
ES-4 Event Constructs	Configuration	– Specifying an event source – Specifying event filtering rules – Specifying event source mapping
VS-5 API GW Constructs	Composition	– Validating API headers, body, and query string parameters – Parsing requests and routing them to the directed backend endpoint – Transforming request and response data to override payload, etc. – Emitting data changes

Operational View

This section shows an infrastructure view to support runtime operations. Since the infrastructure such as racks and power and networking are indistinguishable in solution architectures, A-ESA's infrastructure considerations lean toward technical operations.

Big data processing requires strong infrastructure support. Figure 2-73 shows an immutable infrastructure[21] architecture that supports big data operations. Although the technical infrastructure is *relatively independent* of the application or big data environment, deployment requirements are likely to change, and hardware lead time may result in technology changes.

---

[21] Once a server is provisioned and configured, it generally does not change.

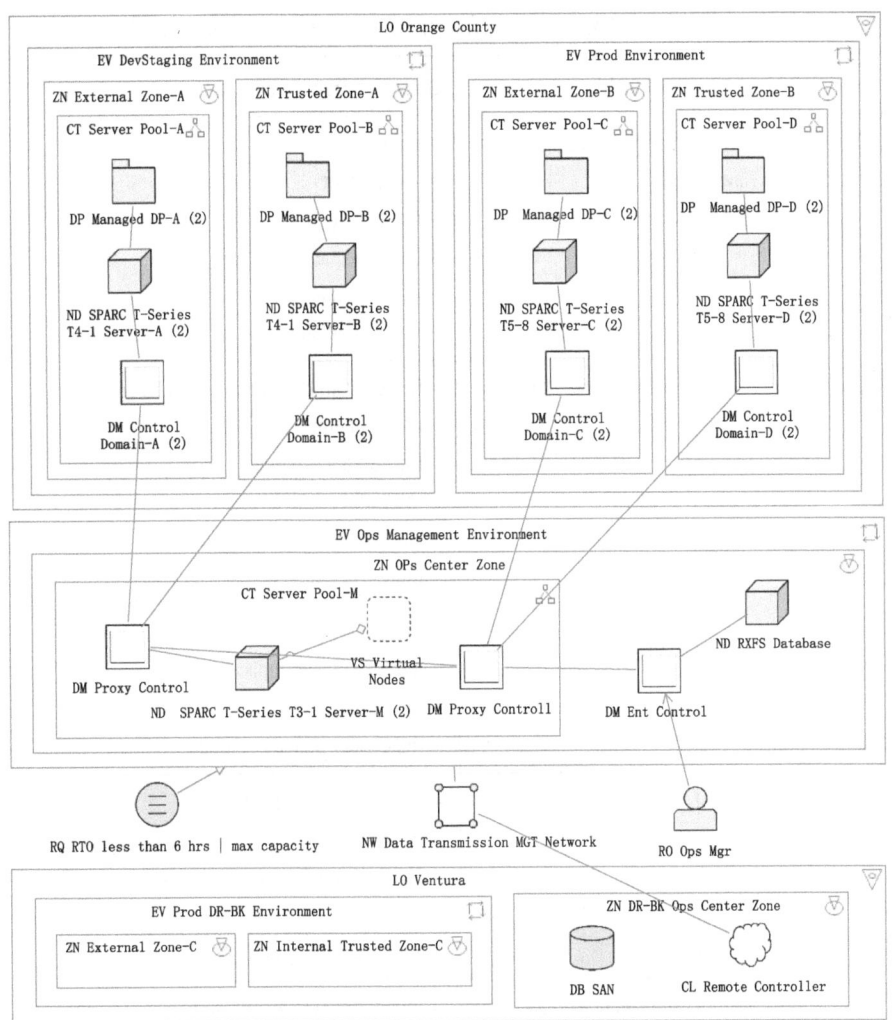

Figure 2-73. *DEP-4 deployment infrastructure architecture view*

Table 2-40 shows property descriptions considered for the runtime deployment environment hosted on Figure 2-73.

Table 2-40. *Properties of Runtime Deployment*

View	Property	Selection Description
DEP-4 Deployment Infrastructure View	Service Level Indicator (SLI)	– Latency – Throughput – Availability – Failed request count – Data freshness – Task-completion rate
	Back up Spectrum	– Sorted from highest to lowest RTO/RPO requirement: – Backup and restore – Pilot light – Warm standby – Multi-site active-active
	Replication Options	– Array-based replication – Network-based replication – Host-based replication – Hypervisor-based replication
	Monitoring	– Application monitoring – Log monitoring – Platform monitoring – Security monitoring – Infrastructure monitoring
	Scalability Factors	– Number of active end users – Concurrent end users – Business transaction volumes – Expected increase of end users – Access mechanisms and locations of end users – Application capacity in each location – Server, network capacity based on the application loads in all locations

Table 2-41 shows the capacity description of the operational environment, of which big data occupies one sixth percent of the total nodes.

Table 2-41. *Properties of Operational Capacity*

Element	Property	Description
DM-Computing	Bare Metal	100
	VM	1500
	Node	240
DM-Storage	Flash Array Storage	– Volume: 80TB (40TB for DR) – Device: 2
	Distributed Storage	– Volume: 180TB (90 HDD + 90 SSD) – Device: 36 (18 HDD + 18 SSD)
	Distributed Object Storage	– Volume: 90TB – Device: 15
	Backup Storage	Volume: 120TB
DM-Big Data	Node	40
Note: – Average resource requirements for a solution production environment: CPU=12 CPU units, memory=16GB, disk storage=500GB, and memory per server=36GB.		

HINT

Role of A-ESA in the IT Infrastructure: A-ESA provides an IT infrastructure (physical facilities and network) architecture blueprint, models hardware into modular components, and has a regular cadence for updating the infrastructure environment as needed, but the infrastructural design is relatively independent of A-ESA.

Model Thinking Questions

Each of the eight example profiles has its own preferred style and solution stage. None of these profiles is intended to be a complete solution profile; rather, each demonstrates a typical use of A-ESA notations with a particular focus, while omitting some others.

The power of A-ESA modeling is not to get it right the first time, but rather to trigger feedback (implicit requirements, inspiration, and the like) and improve the current state of the significant case model in simplicity, clarity and holisticity through iterations. The iterations continue until the model agreed upon by key stakeholders and proven architecturally acceptable within constraints through validation.

One of the most dangerous phrases is: "We've always done it that way," as Grace Hopper said. So, when applying A-ESA model, always think architecturally for architectural fitness, level of abstraction and focus, appropriate reflection of viewpoints, tradeoff exploration, and walkthrough validations.

For better enterprise solution modeling, the following includes a list of architectural thinking questions for each of the example profiles. Some of the questions are conceptual or seemingly obvious, others can be automated by tools, and still others are thought-provoking. For example, those in terms of coupling, cohesion, layering, and isolation at the IT solution level, which in turn effectively determines the solution quality and solution sustainability.

HINT

Architectural vs. Logical Thinking: Consultants are generally good at logical thinking through linear or tree-like analysis and critical reasoning using the MECE (mutually exclusive, collectively exhaustive) approach. Architectural thinking goes beyond this to include systems thinking for structural and dynamic discovery and high-level design with modeling support and tradeoff considerations.

Questions on Profile 1: Strategic Architecture

1. (Figures 2-1 and 2-2): What are the key business priorities (reliability, agility, innovation, etc.) based on the principles in the profile?

2. (Figure 2-2): Is the guiding principle "Use proven and mature technologies" necessary, since it only points to a single solution principle?

3. (Figure 2-2): How will the heatmap analysis be performed for these capabilities?

4. (Figure 2-3): Will this capability view be part of the business architecture?

5. (Figure 2-3): Based on the capability view, is the organization likely to be a functional, matrix, or a hybrid structure?

6. (Figure 2-3): How could the capability view be leveraged and adapted to the needs of the mobile order management system?

7. (Figure 2-3): How does the A-ESA model cross-check missing or repetitive capabilities to ensure they are consistently named and referenced?

8. (Figure 2-4): How does the cloud capability consider people, process, and technology?

9. How and at what approximate level are the capabilities and their corresponding IT services appropriately aligned?

10. Must each capability have sole ownership? What is team ownership?

11. How would the principles be expressed in the profile as a shared governance?

12. How do you track these principles in model?

13. What is the difference between principle, policy, rule, and guideline?

14. What is the difference between principle and governance?

15. How does the enterprise principle element relate to the key choice consideration or architecture decision?

16. What is the IT strategy for handling operational data volume burst?

17. Why are visual representations important for strategic architectures?

18. How should shared capabilities (or services) be designed for business logic containment?

Questions on Profile 2: Lean Business Architecture

1. (Figure 2-5): Should the view elements in this pattern be correlated in the model?

2. (Figure 2-5): For the requirement model, how will the IT services be structured and correlated for significant case scenarios and KPIs?

3. (Figure 2-5): For the interface model, what are the key interface collaborations and their application considerations? What are the significant operations and exceptions?

4.   (Figure 2-5): For the functional service model (in conjunction with the application architecture), what are the focus areas composed of analytical subject areas and measure data elements, and what are the metrics that impact data integrity, distribution, or federation?

5.   (Figure 2-6): Who is the audience for this use case? Business users, architects, solution designers, or developers?

6.   (Figure 2-6): What are the responsibilities of each user case service?

7.   (Figure 2-6): Is this not a significant requirement?

8.   (Figure 2-6): What is the difference between a business use case and a system use case?

9.   (Figure 2-7): How does the capability model relate to the business process model?

10. (Figure 2-7): Is it a significant requirement?

11. (Figure 2-7): Is the human role of each task clearly expressed?

12. Where should usability requirements be defined and illustrated?

13. Can the functional service granularity be defined in business architecture or business analysis?

14. How is a business service reflected in the A-ESA model?

15. What's the difference between actor and role? What's the difference between actor and user?

16. How would you create a view to express the demarcation between human tasks and automated tasks?

17. How does a data service differ from data object, data entity, and data subject area?

18. How does the interaction or collaboration view relate to the use case?

19. A-ESA does not contain a goal element. How might a goal statement be expressed in the model?

20. What are the business priorities in evaluating microservice architecture other than innovation, efficiency, and reliability?

Questions on Profile 3: Reference Architecture

1. (Figure 2-13): How can a business-centric reference architecture serve as a solution architecture overview?

2. (Figure 2-14): Does it contain a common lexicon and taxonomy for data analytics?

3. (Figure 2-14): How could data service be expressed in other A-ESA elements in terms of data service interface, data domain, data caching, and data integrity?

4. (Figure 2-16): What are the justifiable reasons for adopting cell-based architecture other than scalability?

5. (Figure 2-16): What are the limitations of cell-based architecture?

6. What is the difference between reference architecture and architecture pattern?

7. Can reference architecture be used to enforce architectural conformance?

8.   Is it advisable to use reference architecture without a large user community or extended proven cases?

9.   Which A-ESA operational service element is rarely expressed in reference architecture?

10.   Is it possible to make a quality prediction about a solution system based solely on its architectural appraisal?

11.   Will the same architecture result in different development costs?

12.   Does an architectural style impose constraints on the solution design?

13.   What are the key considerations for security reference architecture?

14.   Layering is vital and powerful in an architectural pattern. Can you give examples of bad layering?

15.   What will an SOA operating model look like?

Questions on Profile 4: Integration Architecture

1.   (Figures 2-19 and 2-20): How could the capability views be rendered differently in the solution profile?

2.   (Figures 2-19 and 2-20): Does the solution integration architecture include all the capabilities?

3.   (Figure 2-22): What is the difference between two UC Queue Management elements in this use case? Is such a naming advisable?

4.   (Figure 2-22): What is the real ESA meaning of the element in the GP prefix?

5.   (Figure 2-23): Is it a business process, an application process, or a technical process?

6.   (Figure 2-24): How will these elements be connected judging from other views?

7.   (Figure 2-24): What is the governing idea of this architecture outline view?

8.   (Figure 2-24): How can the principle elements be more specifically associated with domains or elements?

9.   (Figure 2-24): What is the likely audience and technical level?

10. (Figure 2-26): What are the generic services for?

11. (Figure 2-26): Does this view cover all the metrics for this solution?

12. (Figure 2-26): How are the enterprise strategies and goals reflected in the metrics elements?

13. (Figure 2-26): In the context of this architecture, what is the big architecture decision that is not mentioned in the figure?

14. (Figure 2-29): Is the DP Protocol Wrapper a runtime deployment package element or a design package?

15. (Figure 2-29): Does the view express clear interface requirements for the message?

16. (Figure 2-29): ViewFrame is a powerful element that contains a subset view. What if there is no such an element in the face of large and complex solution architecture?

17. Does reusability depend on the complexity of inter-service communication and coordination?

18. What are the primary NFR concerns of this solution? How are they handled in the model?

Questions on Profile 5: Application Architecture

1. (Figure 2-32): What is the essential message conveyed by this architecture overview? Is it necessary to specify element relationships in this view?

2. (Figure 2-32): Functional services are divided into application services (app logic and data services) and technical services. Why? When are they not needed?

3. (Figure 2-32): Is layering properly defined? Are the application services separate from the technical implementation?

4. (Figure 2-32): Is the PS Order Process an atomic, composite, or process service?

5. (Figure 2-33): Who determines the RACI[22] responsibilities in the governance element?

6. (Figures 2-34 and 2-35): How are the data distribution logics accounted for?

7. (Figure 2-36): Is the idempotent consumer pattern better expressed as an interaction view?

8. (Figure 2-41): How are the functional services lumped together by their SLC?

---

[22] A project management acronym for the different responsibility types: Responsible, Accountable, Consulted, and Informed.

9.   (Figure 2-45): Which of the functional services are independently deployable with their own private database schema? How is the data schema related to the bounded context?

10. (Figure 2-47): What are the architectural issues in this as-is view that need to be verified or resolved?

11. (Figure 2-47): Where are the hotspots in terms of key non-functional requirements?

12. (Figure 2-48): What would be the impact of a critical failure of the payment interface? How do you define a critical failure?

13. (Figure 2-48): How can the security considerations (confidentiality, integrity, non-repudiation, and access control) of the payment interface be improved? What are the security domains and zones?

14. (Figure 2-48): What is the depth level of the validation view?

15. Where are the caching data considerations? How are they handled?

16. How does the architecture outline view differ from the functional service view or operational service view?

17. What are the business services in order management systems that can be easily automated as microservices?

18. What is the difference between architectural interaction/collaboration view and architectural relationship view?

19. Why does A-ESA advocate using relationship views instead of interaction views?

20. The solution does not present the critical business processes. What would be the critical processes in an order management system?

21. What are the differences between application architecture and application design in terms of abstraction, granularity, layering, coupling, and isolation?

22. How could gradual system degradation such as multiple process instance data bottlenecks be prevented through modeling?

23. How does the use case relate to the business process?

24. Did you identify any orphan element from the profile?

Questions on Profile 6: Microservice Architecture

1. (Figure 2-51): How is this pattern used in the solution?

2. (Figure 2-56): What is the governance policy for microservice design?

3. (Figure 2-56): Is there a decision on the use of telemetry technology to monitor system health data in the cloud environment?

4. (Figure 2-56): Is there any guidance or standard referenced in the architecture decision?

5. (Figure 2-56): What are the tradeoffs of these decisions?

6. (Figure 2-56): What are the key consideration factors when choosing cloud middleware?

7. (Table 2-28): Where can the circuit-breaker be used?

8. (Table 2-28): How do you associate with the various API endpoints with their service groups?

9. (Figure 2-57): Is this pattern clear enough to understand?

10. (Figure 2-59): What else does the operational service view need to specify for the availability SLA?

11. (Figure 2-59): Do all these microservices have to be deployed all together? Or does each microservice adhere to the independently deployable principle?

12. How are stability, reliability, scalability, and fault-tolerance considerations addressed in this solution?

13. Does the solution modeling process include risk identification and minimization of architectural and technical debt?

14. How is the widely accepted 12+ factor application principles[23] applied to the design of adaptive microservices?

15. Assuming a DevOps view is required, what are the processes and tools to support CI/CD?

16. If you maintain different deployment environments (test, production, or staging, QA), how do you present a release propagation view in a clear sequence and coherence?

17. How could you clearly express SLA feasibility within the budget constraint?

---

[23] 12+ factor application principles include a single codebase, exposed via an addressable URL, API first, isolated dependencies, isolated port binding, in-process execution, concurrency, disposable, build/release/run, dev/prod parity, externalized configurations, one-off admin processes, logging, telemetry, and authentication/authorization.

18. Microservices face many security challenges. In which view do you specify middleware tools such as static application security testing (SAST) and dynamic application security testing (DAST)?

19. How are the services exposed to clients? Is the API versioning approach specified? How is event versioning accounted for?

20. What element properties are required to run a TPS simulation?

21. Is this a cloud-native solution? Which element should be used to enforce the viability of this solution?

22. Will the distributed microservice architecture fit into an enterprise resource planning solution or a data integrity compliant solution?

Questions on Profile 7: Cloud Architecture

1. (Table 2-36): How do you handle events raised by those child services?

2. (Figure 2-60): How do you map the cloud capabilities to the FaaS application solution?

3. (Figure 2-61): Is it too granular to treat construct service as capability?

4. (Figure 2-62): Judging from the case scenario, is it reasonable to put all functional services into a single deployment package?

5. (Figure 2-63): Are the risks adequately mitigated by guardrails?

6.   (Figure 2-63): Do RK elements represent constraints, risks, or issues?

7.   (Figure 2-64): Does it reflect the DevOps toolchain for the solution?

8.   (Figure 2-65): Are the event-driven relationships clear?

9.   (Figure 2-66): What are the remote system management concerns?

10.   If you were asked to create a solution system context view, what would be the control flow and information flow between the to-be solution and its interacting external actors?

11.   How do you express the effort for in-house development vs. cloud hosting?

12.   Does the solution reflect a zero-resource (hot/cold start) serverless architecture?

13.   What is the downside of cloud hosting?

14.   What is the best modeling approach to express the return on investment (cost, time, and quality) associated with the solution?

15.   How do you express the future requirement considerations in a view?

16.   How could you model a large-scale solution (refer to Chapter 5, "Large-Scale Website Architecture") in a cloud environment using A-ESA?

17.   Is this cloud architecture also event-driven architecture, microservices architecture, or an Agile architecture?

18.   What is the access policy and ownership of the data in the cloud environment?

19.   Which A-ESA elements play a critical role in cloud-based reliability and high availability?

Questions on Profile 8: Big Data Technical Architecture

1.   (Figure 2-67): Any major big data capabilities missing like Apache Kafka?

2.   (Figure 2-68): How do you clearly express control flow (such as protocol) and information flow?

3.   (Figure 2-68): What are the main NFR concerns in this view?

4.   (Figure 2-69): When selecting a middleware product solution, do you follow a more standard approach through a list of criteria such as organization requirements, product popularity, usage, supportability, compatibility, testability, stability, extensibility, localization support, learning curve, and so on?

5.   (Figure 2-73): Is it a logical view or a physical view?

6.   (Figure 2-73): Why is the network element usually not detailed in the solution reference?

7.   (Figure 2-73): How is the disaster recovery (DR) plan reflected in the view?

8.   (Figure 2-73): If part of the infrastructure is not reliable, can the system still behave as highly available?

9. Who is the likely audience for this set of architecture views?

10. Which A-ESA view should be used to visually describe Infrastructure as Code (IaC) activities between the software platform and the infrastructure?

11. How does the operational view correlate to the data relationship view?

12. Is reliability a runtime quality attribute? How is reliability measured?

13. Do performance, reliability, and availability play an equal role in the enterprise business?

14. Does a DevOps platform require Infrastructure as Code (IaC)?

15. How do you ensure that an IT infrastructure can scale with business growth?

16. What are your responsibilities as an enterprise solution architect in network design and optimization beyond those of a network specialist or engineer?

17. What is elasticity versus scalability?

Summary

The eight typical enterprise solution profiles in this chapter have shown how A-ESA can be applied to different solution environments. You can now use the A-ESA model to build your own solution architecture, whether it is a business architecture, an operational architecture, a security architecture, or any other style.

Note that judging whether an A-ESA model is sound depends on your modeling skills and the problems you are trying to solve. You can't have an ideal model if you don't know what your basic architecture looks like or what you're trying to achieve. For example, if your primary architectural focus is security in an MSA, the model will certainly reflect all of the key architectural considerations in that area, based on reference architectures and related expertise, either in a selected set of views, significant case walkthrough views, capability views, solution design guidance views, and so on. Also note that the *level of detail* of the model depends on your business needs and solution complexity and is incremented in an Agile approach.

In practice, some individuals only think of high-availability solutions as IT architecture, and some people tend to view ESAs primarily from a technical standpoint, which can limit their understanding of the broader implications for the IT architecture. When this is the case, it is recommended that an ESA be modeled in an S3 approach to ensure its long-term sustainability. Certainly, a solution profile model such as a technically driven architecture has a place in A-ESA.

The next chapter provides food for thought about the architectural thinking process and show you how to form a solution model through an A-ESA thinking framework that ensures a holistic model for a sound ESA.

PART II

A-ESA Governing Ideas

CHAPTER 3

The A-ESA Thinking Framework

Architectural thinking is the meat and structural modeling the bone.

Chapter 2 walked you through example solution models. Now that you have a taste of A-ESA modeling, this chapter introduces the A-ESA's architectural (model) thinking framework, which guides the modeling work with ease and methodological assurance.

An enterprise solution architect generally needs to be good at IT strategic thinking, IT structural thinking, and decision thinking. A large and complex solution requires centralized thinking. However, it's hard or virtually impossible to validate or think most effectively without a sound modeling approach. Modeling is a systems or holistic thinking process, much more than just eye candy. The modeling process is a way of working and a way of thinking. A qualified ESA architect is a thinker and a modeler in one person.

© Sean (Chunhong) Gu 2024
S. Gu, *Mastering Enterprise Solution Modeling*,
https://doi.org/10.1007/979-8-8688-0992-7_3

The Distinct Nature of A-ESA Thinking

Based on the common definitions, the IT architecture is made up of three ingredients: *structure, relationship,* and *guidance.* The A-ESA thinking mechanism more accurately reflects this definition and intuitively incorporates holistic relationships and architectural governance into the modeling process.

Unlike the traditional enterprise architecture and IT strategic plan, A-ESA doesn't cover IT initiatives and the business operating model. However, all of the key initiatives and issues will be reflected in the *enterprise capabilities* and *metrics* elements of A-ESA. Those that can't be mapped to A-ESA are treated as IT fluff.

A-ESA represents SA[1] (*building* purpose) along with EA (*planning* purpose) considerations. It should be noted that in an organization of limited size, the distinctions between EA and ESA are not particularly relevant.

A-ESA maps problem space and tackles solution space. In the meantime, A-ESA covers the middle ground between solution architecture and solution design, connecting both ends for solution realization.

A-ESA takes a *service-based approach,* treating service capabilities, IT, or digital services as the main architectural content. It fits various flexible architectures, including service-oriented architecture (SOA), microservice architecture (MSA), event-driven architecture, process-driven architecture, and so on.

Notwithstanding the various arguments about the differences between EA and ESA, between architecture and design, and between SOA and MSA, A-ESA thinking is centered on three principles (or on the S3 principle), as outlined in Chapter 1:

- **Simple:** Pursuing a high-level black-box effect that is easy to follow.

---

[1] The term "solution architecture" is frequently employed to describe a technical blueprint of a solution that meets a specific set of requirements.

- **Significant:** Emphasizing critical cases and issues that matter most to the enterprise, with walkthrough validation.

- **Systematic:** Viewing the system as a select whole from cross-cutting aspects.

In other words, A-ESA represents a concise, concerned, and correlated model. It can take distinctive forms, but any highbrow, detailed, or lopsided model without S3 in mind would defeat the purpose. Any one-dimensional or sporadic thinking, like a wild goose chase, will most often lead to a blind alley.

Remember, when it comes to enterprise solution thinking, as the saying goes, if you aim high, you might just hit the middle, but if you aim low, you only get the bottom. So if you aim high with S3's powerful modeling, you get a better solution. Otherwise, you are likely to end up with architectural debt in the long run.

The A-ESA Architectural Process

Agile ESA adopts simple modeling *steps* that are treated the same as architecture *areas* defined in Chapter 1. In general, there are six steps: 1) Enterprise Capability Considerations, 2) Case Scenario Analysis, 3) Architecture Overview, 4) Functional Relationship, 5) Operational Deployment, and 6) Validation. Among them, there are several steps derived from the architecture overview (see Table 1-1): metrics mapping, DevOps environment, migration, integration, and so on, based on the solution nature.

Since the ESA architectural process is iterative and incremental, there is no specification of a strict process sequence. The key is to have a set of correlated model views that reflect all six areas or steps.

In fact, the focal A-ESA thinking activities have already uniquely incorporated each basic element of the A-ESA, as described in Table 3-1. Note that A-ESA thinking is more than system thinking, and an ideal A-ESA is not a complete digital twin of the system, but rather a reflection of the S3 model thinking.

Table 3-1. A-ESA Thinking Coverage

Thinking Activity	Description	A-ESA Notation
Strategic thinking	Principle, capability heatmaps, etc., for cost-effectiveness	[PR], [CAP]
Holistic thinking	Overview for scalability, flexibility	[DM], [GP], [VF], [AOV]
Innovative thinking	Valuable capability, value stream, future use case, etc., for adaptability	[VL], [CP], [UC]
Stakeholder management	Stakeholder responsibility, communication, etc., for different viewpoints	[RO], [ORG], [RQ], [TK]
IT and business alignment	Business vs. application services for significant case scenarios	[UCM], [PRM], [PFV], [GS]
Structural thinking	Functional and operational services for composability, isolation, or service orientation	[UI], [AS], [DS], [TS], [SI], [SC], [DP], [MW], [ND], [SRV], [DEP]
System thinking	Architectural integration for interoperability and dynamic context	[SY], [NW], [LO], [SIV], [IRV], [DPM]
Continuous delivery	DevOps environment	[DEV], [SCR], [AR], [DL]
Critical and comprehensive reasoning	Walkthrough and metrics mapping for consistency, verification, and validation	[VLD], [MTS]
Tradeoff thinking	Key architectural considerations	[KC], [ML]
Governance and standard	Governance, reference architecture, and pattern	[GV], [PTN]
Risk management	Architectural and technical debt, risk, and constraint	[RK]

Note:

– Different ways of thinking overlap; for example, structural thinking is a part of systems thinking.

You can use an informal four-step process cycle: SMVG (sketch, model, validate, and govern) to grow an A-ESA model. Surely, you can also use your familiar framework, method, process, or activity list to map an A-ESA model. Table 3-2 illustrates an EA Activity List mapping with the A-ESA notations. Table 3-3 exemplifies the mapping between the 5W1H method and A-ESA.

Table 3-2. *EA Activity List Mapping with A-ESA Notations*

EA Consideration	Example A-ESA	Concern
Stakeholder	RO, ORG	Interest, trait, cooperation, responsibility, decision-making, etc.
Business Objective	VL, PR, RK, CAP, AOV	Mission, vision, problem space, pain point, constraint, etc.
Solution Objective	PR, RK, DM, DL, AOV	Solution scope and context, future case, interoperability, etc.
Requirement Needs	RQ, UC, PRM	Non-functional requirement, future requirement and *implicit requirement*, etc.
Building Blocks	GS, DM, VF, SRV, DPM, CAP, AOV	Level of decomposition and building block, etc.
Asset MGT	DL, GV, KC, PTN, CAP	Reusable asset, etc.
Review	KC, RK, DL, VF, VLD	Iterative review, gap analysis, etc.
Action Plan	GV, TK, AOV	Solution management, governance action, etc.

Note:
– Depending on the solution environment and the A-ESA adoption process, one or more of these notations represents each consideration.

Table 3-3. *Mapping Between the 5W1H Method and A-ESA*

5W1H	A-ESA Element	A-ESA View
Who	Role	Organization View
What	Task, Use Case, General Service, Domain, Requirement, Value, Risk, Principle	Case Scenario View, Capability View
When	View Frame, Deliverable, Future Requirement or Future Use Case	As-Is and To Be Views
Where	Location, Domain	Operational Deployment View
Why	Key Consideration, Risk, Governance	Metrics View
How	Functional and Operational Services	Collaboration Relationship View, Deployment Package Mapping View, Walk-Through Validation View, RA/Pattern View

As you can see from Tables 3-2 and 3-3, A-ESA already *inherently* includes elements and views that are specified in many traditional architectural processes and methods. By applying the base set of A-ESA elements and views, you will naturally cover all enterprise solution process activities. The pending or flagged A-ESA notations (views, elements, and properties) can serve as a constant reminder of related activities.

The problem with many seemingly good architecture frameworks, methods, processes, or guidelines is that they don't necessarily specify all the key elements and views like A-ESA does in a simple, related, and proven modeling approach. In contrast, A-ESA's thinking framework is *model-bound* and effective for solution landing.

Pillars of the A-ESA Thinking Framework

The uniqueness of the A-ESA thinking framework is reflected in three pillars (see Figure 3-1). These three pillars can be broadly expressed as two traditional schools of architectural thought: *decisional architecture* and *structural architecture*, plus *landscape view architecture,* which outlines solutions at a higher level of thought. The three pillars are intertwined and influence each other, so all three architectures must be well correlated and synchronized. In addition, the three architectures are inherently subject to the cost/benefit considerations of the solution deliverable.

For a solution architecture that is not that large, the decisional architecture and the structural architecture suffice. However, for overly complex and large-scale solutions, the landscape view architecture plays a critical role.

The three pillars in A-ESA are supported by 1) abstraction and focal coverage, 2) system aspect, and 3) metrics (see Table 3-4).

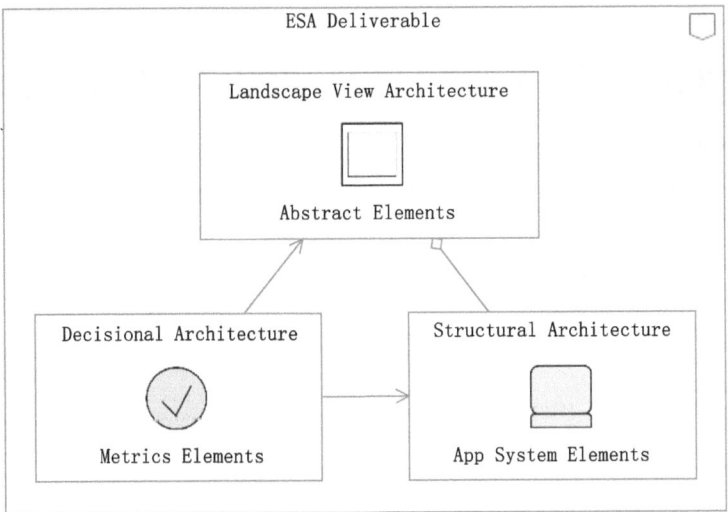

Figure 3-1. *The three pillars of the A-ESA thinking framework*

Table 3-4. *Model Embodiment of A-ESA Thinking Framework*

Model\Thinking	Abstraction	System Aspect	Metrics
Views	Outline Views	System Arch Views	Metric Mapping Views
Elements	Abstract Elements	System Arch Services	Metric Elements

Thinking of the famous *known-unknown matrix,* the A-ESA architectural thinking is meant to discover unknown information or solve unknown issues, whether they are implicit requirements or unforeseeable conditions. The A-ESA framework is intended to guide a better *servitization* process and address architectural issues that are not easily detectable at the design level, and in the meantime, it does not delve into data structure design, application design, infrastructure design, and IT service management.

The next section includes a brief description of the A-ESA thinking framework that helps shape a good IT model.

Abstraction for Landscape View Architecture

A-ESA helps create an IT landscape or *holistic* architecture that supports different levels of abstraction using the architecture elements in Table 3-5.

Table 3-5. *Abstraction Elements*

Category	Element	Description	Containment
Scoping	View frame	A representation of views for architecture view scaling and drill-down mapping	Strong
	Domain element	A composition element for physical grouping with elements	Strong
Grouping including layering and partitioning	Group element	A common element for generic, logical, or conceptual grouping without strict containing relationship	Weak
	Domain element	A composition element for logical grouping with elements	Weak
Service granularity	Generic Service element	A generic service that represents business service, conceptual service, macro-service, mini-service, and the like	Weak
	Process Service element	A process or composite service that contains child elements in either a choreography or orchestration scenario	Strong
	Virtual Service	A service that does not exist in a clear physical service interface, form or shape, but provides service interactions	Weak

Note:

– *Domain* in ESA often refers to a physical grouping. However, domain can also denote a logical domain or business domain. A domain may have an overlapping relationship with other abstractions. So, use this term with caution in different contexts for clarity.

Just like zooming in or out on a map, an ESA architect chooses the right *level* and *dimension* of abstraction through model thinking and manipulation. You can use the standard set of A-ESA notations for a recommended ESA abstraction. Or you can selectively use base elements plus assistive elements or your own special images that map to base elements for the level of abstraction that fits your solution environment.

For a mega-sized enterprise solution architecture, a layered responsibility approach could be adopted. For example, the *chief architect* is responsible for the overall enterprise solution model more at the EA level, *lead architects* for the segmental solution models, and *developer teams* for various architectural design models. All these models at different levels follow the A-ESA's *S3* objectives (*simple, significant,* and *systematic*) to avoid unfocused modeling, and they should be correlated for holistic view considerations.

A well-architected complex system is usually made up of many stupidly simple systems, and a simpler abstraction often comes from a detailed mess. A critical task of an A-ESA architect is to determine both the *layered level* and the *granular levels* of system abstraction to better integrate different *dimensions* (services, technologies, management, governance, etc.) and different *domains* (application, data, security, etc.). Architectural abstraction promotes *holistic* thinking beyond the scope of solution assurance. A large-scale enterprise solution architecture without a clear landscape view architecture is bound to fail, or it will not be viable in the long run when substantial changes take place.

System Aspect-orientation for Structural Architecture

Now you come to the *Solution Operating Model (SOM)*, or Solution System Model, which is part of the A-ESA model that includes functional and operational services. The system model consists of three dimensions that underscore cross-cutting considerations.

Figure 3-2 shows the cross-cutting aspects of the A-ESA SOM, where functional and operational services across application, data information, and technical levels as well as non-functional aspects such as reliability, performance, security, and so on.

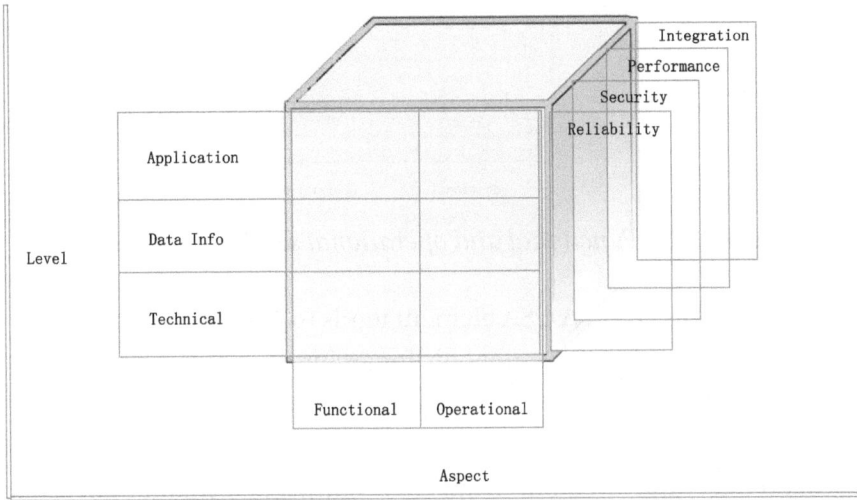

Figure 3-2. *A-ESA SOM for enterprise systems*

Functional and operational services act as the glue at different levels and for different aspects, as shown in Figure 3-3. For functional services, the lower the level, the more technical the service. Similarly, for operational services, the lower the level, the more technological or base platform-specific the service.

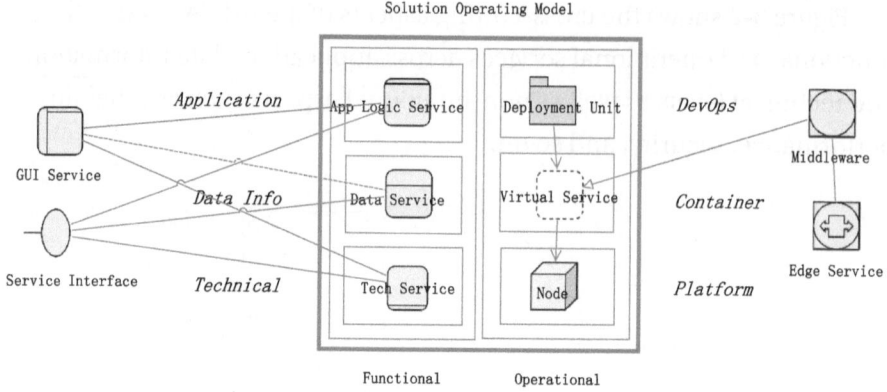

Figure 3-3. *Functional and operational services in the SOM*

Because each type of A-ESA element tends to have the same *service-level characteristics,* it's better not only for architectural integration, but also for architectural adaptability and other cross-cutting hotspot concerns.

For architectural integration and cross-cutting considerations related to the four basic EA architecture styles (BA, AA, DA, and TA), the A-ESA elements in the Solution Operating Model perform the following:

- The **User Interface (UI)** element links business architecture via requirement case scenarios primarily with application architecture and technical architecture.

- The **Service Interface (SI)** element exposes functional services to application programming interfaces.

- The **App Logic Service (AS)** element transitions between business architecture and application architecture, uses technical architecture, and optionally collaborates with data services.

- The **Data Service (DS)** element provides data services to other architectures.

- The **Tech Service (TS)** element supports application and data architectures.

- The **Deployment Package (DP)** element represents a deployment unit that may contain different sets of functional services for SLAs considerations. It also bridges the transition gap from the application and data architectures to the technical architecture.

- The **Virtual Service (VS)** element represents containment for runtime deployment packages. While deployment packages are considered more from a performance aspect, and virtual nodes are considered more from a high-availability aspect, they are interdependent.

- The **Node (ND)** element represents technology foundations that are agnostic to the application environment.

- The **Edge Interface (EG)** element represents the edge interface or adaptation agent used to connect middleware and system services.

A complex system comes with complex case scenarios, but many system complexities are due to architectural complexity from entangled component design and implementation. As a rule of thumb, the solution space should be no more complex than the problem space. It is observed that a structural architecture without a service-based approach can lead to unnecessary solution complexity.

The SOM model helps you think through iteratively with systems thinking and inherent SLC-matching isolations. It can be considered along with appropriate measurement metrics, a workable architectural style, and the right governance techniques. In short, don't just follow a trendy solution architecture or a solution-dependent success story, but stick to the foundation of systematic service-based architecture.

Metrics Mapping for Decisional Architecture

IT architects are usually pretty good at structural architecture, but some of them often overlook decisional architecture, which requires a mix of art and science and is more challenging to learn. The former helps you identify the *hotspots* that need the most attention, while the latter makes sure you're hitting the *sweet spots*, given the constraints.

The decisional architecture is enforced by the A-ESA *metrics* elements. As shown in Figure 3-4, these metrics elements represent three levels of architectural importance: enterprise IT strategy, solution assurance, and architectural conformance. Taken together, these three levels *promote holistic thinking* as opposed to traditional systematic thinking from a *solution assurance* perspective only.

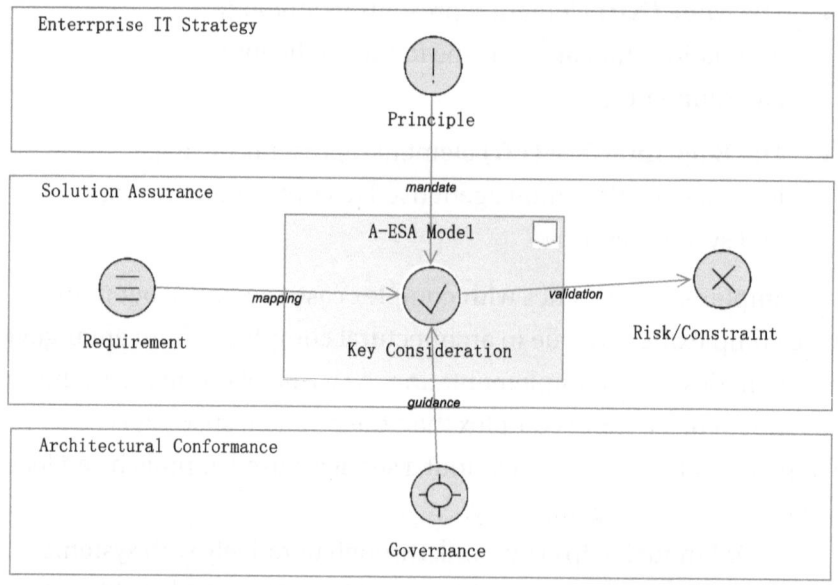

Figure 3-4. *Decisional architecture metrics*

The *key choice* element is influenced by the other metric elements. It's mandated by the principle, based on requirement needs, and constrained by the risk factors and governance rules. IT solution architecture is always a trade-off process within the context and constraints.

Note that these five metric elements influence each other. A slightly different mapping of the relationship between these elements is likely. However, missing any one of the elements will likely result in *architectural debt*. For example, without the principle element, you may produce a solution that does not map to the enterprise strategy, or your solution may not work well with other enterprise solutions.

Certainly, metrics mapping has an impact on the choice of architectural style, capability mapping, measurement considerations, and service orientation. For example, metrics mapping, along with the service-oriented architecture methodology, can help determine the business alignment of application services and filter out significant services or processes.

A-ESA Modeling Effects

Along with A-ESA's S3 intent (simplicity, significancy and systematics), the objective of A-ESA modeling is architectural clarity, responsibility, and quality.

Architectural Clarity

Architectural clarity is the foundation of solution assurance and architectural conformance. *Lack of clarity leads to ineffective communication and complexity.* In practice, arguments and misunderstandings are often caused by different viewpoints, backgrounds, and levels of detail. Much technical debt, including architectural debt, is caused by unclear enterprise solution architecture.

Unambiguous Terms

Table 3-6 is an example of some elements that may have different meanings that must be determined in the solution settings. They must be unambiguous and properly categorized within the *context of the solution architecture*. For example, a microservice can be a concept, a software architecture style, a set of best practices, a design approach, an application service, a deployment unit, and so on, and it can be a fuzzy term without a clear architectural specification.

As you know, a microservice or group of microservices is essentially an independently deployable unit. A specified, bounded context that does not match its intended deployment unit will lead to confusion and likely to architectural debt. Therefore, it is important to clarify what an IT concept really means in the ESA context.

Table 3-6. *Examples of Elements Subject to Clarification*

Defined Element	A-ESA Connotation	Description
Microservice	Principle [PR]	Such as single responsibility, independent deployment, lightweight communication.
	Generic Service [GS]	More from conceptual analysis: an initial fitness of a microservice.
	Domain [DM]	More from business analysis: a microservice domain is a specific area of business functionality.
	Bounded Context [BC]	More from application analysis: a logical boundary around a specific domain area.
	Composite Service [AS]	Domain microservice or a choreographed group of service.
	Process Service [PS]	Integration microservice or a group of orchestrated service.
	Data Service [DS]	When a microservice serves as a data layer
	Tech Service [TS]	Unit-of-work (stateless) microservice or supporting service.
	UI Service [UI]	When a microservice serves as special UI functions.
	Microservice [MS]	A small, independent service within a larger application. It can include or be part of an application logic service, data service, technical service, or service API.
	Application [AP]	A suite of small, independent services that work together to form a larger application.
	Deployment Package [DP]	More from an operational aspect: a containerized instance of a microservice, packaged with its dependencies, configuration, and runtime environment.
	Virtual Node [ND]	In runtime, the physical form of a microservice is typically a container or a lightweight virtual machine.

(continued)

Table 3-6. (*continued*)

Defined Element	A-ESA Connotation	Description
Data Service	Database Middleware [DB]	The database package or direct database service.
	Database Service [DS]	The service that accesses the database.
	Data Object [OB]	Intermediate layer data service.
Frontend	Device [MD]	Mobile device.
	Browser [BR]	External user interface.
	GUI [UI]	Application interaction interface.
	PC Device [PC]	System device.
Cloud	Environment [EV]	Virtual environment, Code as Infrastructure environment.
	Platform [PL]	Fee-based PaaS, IaaS platform.
	Container [VS]	Virtualization container, virtual node.
	Interface [SI]	Cloud interface.
Edge Service	Adapter [EG]	Adaptation service.
	API Interface [PI]	Adaptation API.
	Middleware [MW]	Interface via middleware configuration and customization.
	Technical Service [TS]	Business-agnostic interface implementation.

More specifically, architectural clarity can be achieved in several steps, including these:

- Clarify architectural connotations from hype words, business terms, habitual naming, jargon, and so on.

- Clarify case scenario issues using an Agile approach. For example, event storming for the use of Saga patterns for data reconciliation and consistency, and then mapping to the architectural view(s).

- Clarify the intent and result of the solution and create architectural services in between if needed.

- Leverage the A-ESA thinking framework to convey each of your solution elements in an ESA model form, or go back and rethink it.

Enhanced Architectural Communication

A-ESA presents the same issue to different stakeholders with different viewpoints, using the same notations, likely in different rendering forms (cartoon-like image, colored icon, etc.) or modes (simple, assistive, etc.), so everybody understands the issue from the same page. A-ESA's multi-dimensional views of structure, decision, and abstraction ensure eventual correlated consistency between technical, business, and architectural models, or between different architectural layers and the like.

Table 3-7 shows the multiple model modes for element inclusion. Each A-ESA user may only apply and view one model mode in an Agile and effective approach that best fits their model view preference.

As part of model customization, the ESA architect decides how to use different model modes to achieve a *better communication effect*. For example, you might use a primary mode (e.g., base mode) for the development team and an optional lean mode for a target audience.

In A-ESA, the base mode is the foundation of the model. The assigned model authority is responsible for the translation between the base mode and any other modes.

Table 3-7. *Multiple Model Modes*

Model Mode	Purposeful Effect	Primary Audience
Base Mode	For A-ESA compliance and effective architectural modeling	ESA architect
System Mode	For systems modeling containing only functional and operational elements	ESA architect, digital twin agent
Lean Mode	For small, fast, or initial solutions with simplicity in mind	General user, Agile practitioner
Standard Mode	With a commonly used set of assistive elements for better descriptive and visual understanding	Solution stakeholder
Special Mode	With added assistive elements or special images to better communicate with those unfamiliar with A-ESA	Specialist, customer, public audience

Note:
– Model Mode is primarily used for A-ESA elements. For model views, you can use a set of lean model views or a flexible set of views.

Architectural Responsibility

A-ESA is not just for the sake of modeling, but also for *better solution management*. The ESA service definition helps divide solution work for architectural focus, as exemplified in Table 3-8.

Table 3-8. *ESA's Division of Responsibilities*

A-ESA Notation	Division of Work	Architectural Focus
User Interface [UI], Page Flow [PFV], etc.	UI Role/Team	Business Architecture
App Logic Service [AS], Application [AP], Process [PS], etc.	Application Development Role/Team	Application Architecture
Data Service [DS], Data Schema [SM], etc.	Data Management Role/Team	Data Architecture
Tech Service [TS], Deployment Unit [DP], Edge Service [EG], etc.	Technical Role/Team	Technical Deployment Architecture
System [SY], Node [ND], Network [NW], etc.	Infrastructure or Platform Role/Team	Technological Architecture

The division of labor may seem simple, but its determination has a profound impact on solution deliverables. For example, in the real world of solution settings, an improper division of responsibility can often lead to isolation issues, where a UI service contains a lot of business logic, an application logic service contains technical services, or multiple teams handle long-running processes across different organizational groups.

Architectural Quality

This section looks at the impact on architectural quality.

Purposeful Architectural Measurement

One result of A-ESA modeling is that it has significant measurement mappings, which is discussed in the next chapter.

Since most architecture decisions are constrained by the cost factor, this section is a brief description of how cost becomes an integral part of A-ESA model thinking.

Cost Impact

As a constraint, cost is likely to be a common consideration for an enterprise solution. Despite the budget allocation, the ultimate purpose of ESA is still for the cost of change.

Table 3-9 illustrates the major cost consideration elements in an ESA architectural thinking process.

Table 3-9. *Value vs. Cost*

Focus Space	Description	A-ESA
Enterprise Business	Indicated by strategic value proposition	Value [VL]
Enterprise Solution	Impacted by cost optimization and cost of change	Key Consideration [KC]
Solution Deliverable	Demonstrated by the benefits of its features and usability	Deliverable [DL]
Note: — The solution value is measured by benefit minus cost using this formula: value = benefit − cost. — Cost is specified by a view or element property, treated as a *constraint* element, or reflected indirectly from the *value* element.		

Cost optimization is a combination of strategies, leading practices, solution architectural excellence, and tool support. See Table 4-15 in Chapter 4 (solution architecture) and Table 6-34 in Chapter 6 (solution management) for complementary and overlapping concerns.

Enhanced Architectural Conformance

A-ESA acts as an intelligence booster, as it promotes architectural thinking. A-ESA service-based modeling facilitates systems thinking. The intuitive model spec and thinking framework meets practical needs, with its unique features, for example:

- The link between EA and SA ensures long-term sustainability.

- A flexible set of model modes enables effective communication between different stakeholders.

- Transaction or distribution patterns in terms of NFRs are emphasized in scoping elements and architectural integration elements (functional service element, deployment unit, and optional assistive ESA element) to meaningfully connect the dots across levels, aspects, perspectives, or layers.

- Significant case scenario walkthroughs reduce complexity from the consideration of many business and system requirements to the focused mapping of architectural characteristics for flexibility and reusability.

- Architecture-centric approach drives the use of meaningful ESA element terms and proper specification of abstraction levels

- Comparing and incorporating different architectural styles becomes easy with a minimal set of elements common to ESA solutions.

- The architectural pattern view provides visual guidance as an implementable governance.

- The service component realization view provides typical and reusable leading practices for end-to-end landing assurance.

- The validation view is a constant reminder of hotspots and sensitivities.

- Governance techniques are embodied in the model with a guided mapping of element or property specifications.

- Key solution management mappings bring solution context to architectural considerations.

- The *art* part (guidelines, techniques, and tradeoff balancing) and the *science* part (structures and relationships) are modeled into a complete whole.

- The solution profile approach fosters special area scrutiny as well as system scaling.

- ESA tool support eases adoption and compliance.

In short, a clearly represented, rationally partitioned, well correlated and governed model leads to an effective, quality ESA architecture.

Summary

A well-qualified ESA architect is T-shaped, multi-disciplinary, experienced, and good at architectural thinking. The breath and horizon of an architect's thinking determine the quality of the solution. The value of an architect is their ability to see problems from multiple perspectives. Obviously, the more experienced and skilled the architect is, the broader dimension they bring to their architectural thinking. An ESA model should reflect an architect's mindset in terms of architectural characterization.

An ESA architect, like a helmsman equipped with the thinking framework, steers the architectural ship in the right direction. Modeling with a thinking framework is like architecting with an intelligence turbocharger. The framework allows junior architects to catch up and think like seasoned architects.

PART III

A-ESA Essentials

CHAPTER 4

A-ESA Measurement

A solution model should be measurable and validatable.

One of the constituents of A-ESA is architectural measurement. This chapter lists some common measurement criteria and a rough mapping to the A-ESA Notation by core architectural area.

The measurements can come from a variety of perspectives—user, value, process, product, or service-level requirements. While functional requirements and non-functional requirements (NFRs) both affect the architecture, most architectural issues manifest *non-functional characteristics*. You can treat the A-ESA model as *the NFR authority* and deal with any significant cross-cutting concerns (e.g., security) as separate profiles or a set of specific views.

In the real world, non-functional requirements are often ambiguous, incomplete, or missing. Many of the solution requirements are actually *implicit* requirements (mostly NFRs) that are determined through the solution-modeling process. From various viewpoints, architectural metrics help complete and clarify current quality attributes and future changes. It is recommended that the NFRs be incorporated into the architecture model as early as possible.

© Sean (Chunhong) Gu 2024
S. Gu, *Mastering Enterprise Solution Modeling*,
https://doi.org/10.1007/979-8-8688-0992-7_4

A-ESA typically deals with enterprise solutions in a complex system (or system-of-systems) environment and therefore has to deal with a large number of NFRs. NFRs[1] are also sometimes referred to as *Quality Attribute Requirements* (QARs) or *Cross-Functional Requirements* (CFRs). Table 4-1 is a selected list of 100 common NFRs.

Table 4-1. *A Selected List of Common NFRs from Solution Projects*

Common NFRs			
Accessibility	Customizability	Learnability	Reproducibility
Adaptability	Deliverability	Legality	Resilience
Adjustability	Demonstrability	Locatability	Reusability
Aesthetics	Dependability	Maintainability	Rigidity
Affordability	Deployability	Manageability	Robustness
Affordability	Distributability	Materiality	Safety
Agility	Duplicability	Maturity	Scalability
Analyzability	Durability	Modifiability	Security
Applicability	Dynamicity	Modularity	Sensitivity
Applicability	Effectiveness	Mutability	Serviceability
Asynchronicity	Efficiency	Non-repudiation	Simplicity
Auditability	Elasticity	Observability	Sociability
Authenticity	Enhanceability	Opacity	Specificity
Availability	Evolvability	Operability	Stability

(*continued*)

[1] The book uses the term *NFR* to indicate requirements for quality attributes and constraints.

Table 4-1. (*continued*)

Common NFRs			
Capacity	Expandability	Performability	Standardizability
Changeability	Extensibility	Portability	Suitability
Compatibility	Fault-tolerance	Predictability	Supportability
Complexity	Flexibility	Profitability	Susceptibility
Compliance	Fragility	Programmability	Sustainability
Composability	Generality	Realizability	Testability
Comprehensibility	Granularity	Recoverability	Timeliness
Concurrency	Heterogeneity	Recreatability	Traceability
Confidentiality	Immobility	Redundancy	Utilization
Configurability	Installability	Reliability	Variability
Conformity	Integrity	Repeatability	Verifiability
Costliness	Interoperability	Replaceability	Viscosity
Credibility	Latency	Replicability	Volumetrics

There are many categories of NFRs. For example:

- Microsoft's Azure Well Architected Framework (five categories): Cost management, operational excellence, performance efficiency, reliability, and security.

- Hewlett-Packard's FURPS: Functionality, usability, reliability, performance, and supportability.

- Google's DORA[2] (for DevOps): Deployment frequency, lead time for changes, mean time to recover, and change failure rate.

There is no hard and fast rule for categorizing NFRs. What you need is a set of measurements that fit your solution context. Here are a few practical rules of thumb:

- Don't overlook significant NFRs (e.g., those for application logic service that can't be easily changed later).

- Adapt the measurement to a specific environment (e.g., web or mobile application).

- Include only a minimal but accurate representation of your NFRs in the A-ESA model rather than a detailed specification.

- An NFR can be both a functional requirement and a derived requirement. For example:

  - Functional requirement: Implementation of a wrong password lockout function.

  - Non-functional requirement: Maximum allowed password guess is three times.

Figure 4-1 provides a general category scheme for the measurement metrics in this chapter. It includes enterprise input metrics, quality attributes, and constraints.

---

[2] Google's DevOps Research and Assessment

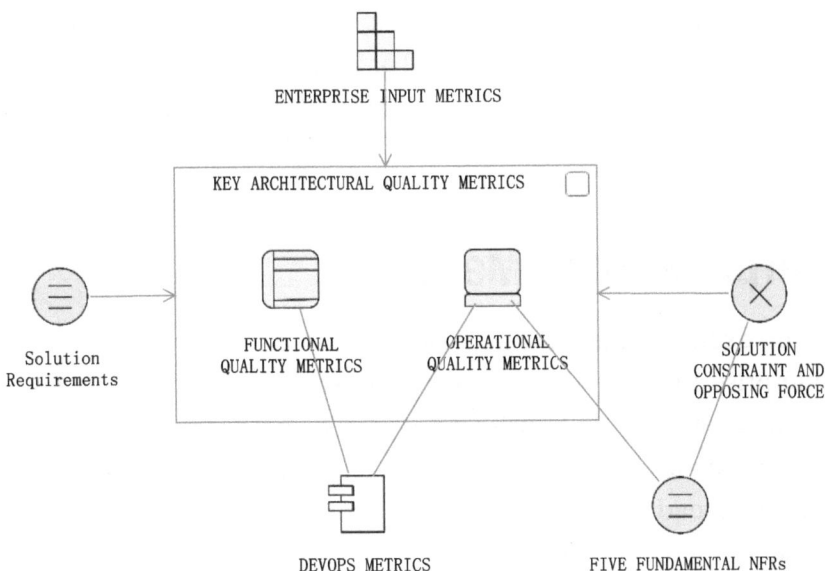

ENTERPRISE INPUT METRICS

KEY ARCHITECTURAL QUALITY METRICS

Solution
Requirements

FUNCTIONAL
QUALITY METRICS

OPERATIONAL
QUALITY METRICS

SOLUTION
CONSTRAINT AND
OPPOSING FORCE

DEVOPS METRICS

FIVE FUNDAMENTAL NFRs

Figure 4-1. *Overall measurement categories*

Note that the A-ESA notation mappings in this chapter are from their typical solution viewpoints, and they are for reference only, because NFR examples are extensive and the meaning of a measurement metric may change from stakeholder to stakeholder, from solution to solution, and from concern to concern. In reality, some solution projects create a set of metrics or objectives in *business terms*. To effectively measure an enterprise solution system, the solution objectives must be clarified, and then the right metrics must be selected to align with those objectives. Modeling can help correlate and analyze metrics on purpose, with the aid of tools.

Remember Goodhart's Law: "When a measure becomes a target, it ceases to be a good measure." Measurement is a means to an end, not the opposite, as is often misunderstood.

For modeling notation abbreviations, see Table 1-2 in Chapter 1 for *view prefixes*, Table 1-3 (Chapter 1) or Table I-1 (Appendix I) for *basic/standard element prefixes*, and Table I-3 (Appendix I) for *assistive element prefixes*.

Enterprise Input Metrics

Enterprise solution architecture is closely related to IT strategic planning or enterprise architecture, which serves as *guidance* input metrics for solution architecture. Enterprise input metrics include strategic guidance such as principles, values, capabilities, governance, constraints, and so on, and can be broad in scope. The following section exemplifies principle and governance input metrics.

Enterprise Principle Guidance

The *architecture principle* (or principle for short) element represents and encompasses all other directional elements, such as goals and drivers, because each architecture principle is indirectly related to business objectives and key architecture drivers. The principle aligns with enterprise strategies and guides architecture decisions.

Table 4-2 shows some *enterprise principle* measurements and their mapping notations (architecture views or elements).

Table 4-2. *Enterprise Principle Measurements*

Example Metric	Example Measurement Description	A-ESA
Guiding Principle	Business-driven, meets the needs of all business models and maintains complete control over key business systems.	[AOV]
Domain Principle	PaaS platform should use the same technology, interfaces, and protocols as other international standards to ensure compatibility and portability.	[SRV] [DEP]
Data Principle	Data is captured once.	[IRV]
Application Principle	Presentation logic, process logic, and business logic are separated.	[SRV] [SIV]
Business Principle	Business flows should be created around value and be scenario based.	[PRM]
Technology Principle	IT requires intelligent operation.	[DEP]

Enterprise Governance Guidance

Many enterprise standards,[3] compliance, and leading practices can be mapped to govern the enterprise solutions. Table 4-3 provides examples of how to map enterprise governance metrics to solution artifact notations.

---

[3] Standards can be de facto, de jure, or even client defined.

Table 4-3. *Enterprise Governance Measurement*

Example Metric	Example Measurement Description	A-ESA
Industry Compliance	Solution must conform to industry-specific healthcare data (HIPAA).	[GV]
Security Mandate	Enterprise system security should be very secure. Firewalls must separate the different layers of the infrastructure.	[NW]
Note:		
– Governance measurement can come from non-functional requirements or as a business constraint. Governance means rules, processes, and leading practices that produce results that meet the requirements.		

Key Architectural Quality Metrics

While the DevOps environment relies heavily on automated testing and monitoring to achieve product or solution quality, it would take many more iterations to detect and fix issues. What is more, some architectural issues, such as granularity and isolation, are not easily detected by tools alone. Good architectural modeling (either up-front or in-process) will eliminate many of the major issues beyond the design and coding implementation. Solution quality issues resulting from architectural design and coding complexity are costly to the business because they are time-consuming to resolve or necessitate a complete redesign.

This section touches on some *qualitative* measurement metrics that are critical to enterprise solutions.

Architectural Structuring Quality

Table 4-4 lists key architectural structuring quality metrics that an enterprise solution architect should spend time on. Their importance can't be overstated.

Table 4-4. Key Architectural Quality Metrics

Metric	Measurement Description	A-ESA
Abstraction	The use of abstraction in architectural design allows for a streamlining of the design process, whereby the most salient aspects of a structure are given prominence while the less significant details are relegated to a lesser position.	[DM] [GP] [GS] [PS] [VF]
Coupling	In the context of architecture, the term "coupling" describes the interdependence between various elements, including systems, domains, modules, services, and architectural types.	[PFV] [PRM] [SIV] [DEP]
Isolation	Isolation is the process of identifying and separating dependencies between services, devices, etc., especially between the application logic and the underlying technology.	[UI] [AS] [DS] [SC] [TS]
Layering	Layering is a way of organizing similar functionalities into groups, modules, or coarse-grained services, with each layer supporting the one above it.	[GP] [PTN]
Tier	Tier is a physical organization of similar operations or devices by layer.	[DM] [DEP]

(continued)

Table 4-4. (*continued*)

Metric	Measurement Description	A-ESA
Conceptual Simplification	Conceptual simplification means reducing architectural complexity through visualization and modeling.	[VF] [AOV]
Structural Simplification	Structural simplification means to simplify architecture through architectural segmentation, partitioning, and structuring, rationalization of information flow and control flow, etc.	[AOV] [SRV]
Service-oriented Granularity	Service-oriented granularity is characterized by loose-coupling, technology independence, and self-containment, and is exposable through standard interfaces. The level of service granularity varies, for example, macroservice,[4] appservice, miniservice, microservice,[5] etc.	[UI] [AS] [DS] [TS] [SI] [MS] [GS]
Deployment Unit Mapping	Deployment unit/package mapping facilitates the transition between the application and the hosting environment.	[DPM]
App Framework Suitability	Application framework suitability determines the application development environment.	[DEV] [KC]
Note: — Some structuring, such as coupling, requires concurrent architectural connectivity considerations.		

---

[4] Like a monolith

[5] Like a group of microservices coming together in a pattern

Architectural Connectivity

Table 4-5 provides a list of integration architectures and their associated mapping notations. These are typically *sensitivity points* that require special attention in the architecture model.

Table 4-5. *Architectural Connectivity*

Metric	Measurement Description	A-ESA
Application Integration	Application integration, commonly called enterprise application integration, is a process that enables the seamless integration of disparate applications within an organization. It is achieved through the use of integration patterns, which can be evaluated through interface compatibility tests.	[TS] [SI] [MW] [AOV] [PTN] [SIV]
UI Integration	UI integration means UI service integration via portals, widgets, API, AIGC (AI-generated content), etc. It can be measured by UI service component plug-and-play capability.	[UI] [PFV]
Process Integration	Process integration is the evolution of operations through business process management, dynamic human resource allocation, workflow automation, compensation, and composite service delivery. It is characterized by the flexibility of business processes, commonly measured by scalability, modifiability, dynamism, and reusability.	[PS] [AS] [RO] [TK] [PRM]
Data Integration	Data integration is the process of combining data from multiple, disparate sources to provide users with a single, unified view. It can be measured by the ability to provide data services through federation, master data management, and data sharing.	[DS] [SI] [MW] [SIV]
Data Flow Integrity	Data flow integrity ensures data and transaction integrity via unambiguous pattern views and data service interactions, or scenario-based simulation.	[DS] [PTN] [IRV]

(continued)

Table 4-5. (*continued*)

Metric	Measurement Description	A-ESA
Architectural Integration	Architectural integration is an end-to-end solution integration. It is generally measured by cross-cutting architectural concerns such as deployment units mapping between application architecture and technical architecture or via DevOps optimal deployment.	[PRM] [AOV] [SRV] [DPM] [DEP]
Interoperability	Interoperability is the ability of different software applications, platform solutions, products, systems, or devices to connect and communicate in a consistent manner. This is accomplished through the use of adapters, administrative shells, and other such tools, with results measured through standardized interface compatibility testing.	[SI]
Compatibility	Compatibility means two things can work well together. It can be measured by the ability to coexist and interoperate via a standardized interface.	[SI]
Composability	Composability is about combining smaller parts to make bigger, more complex systems. It is measured by an architectural modeling with service-based functional services.	[AOV] [SRV]

Note:

– Composability can fall into other categories such as adaptability and maintainability. However, it plays a large role in the integration of enterprise solutions.

Functional Quality Metrics

This section lists functional quality metrics that are often a big part of the Service Level Agreement.

Note that even from a functional perspective, A-ESA focuses on quality attributes, not business requirement features or tasks. As a rule of thumb, *don't let functional requirements drive your architecture*, because IT architecture is primarily about structure, relationships, and guidance.

Business Functional Quality

Table 4-6 identifies the major business functionality quality metrics. They are relatively easy to measure and require little effort in the solution architecture.

Table 4-6. *Business Functional Quality*

Metric	Measurement Description	A-ESA
Functional Appropriateness	The degree to which the software application, product, or system solution is in compliance with the specified tasks and objectives.	[AP] [PD]
Functional Completeness	The degree to which the software application, product, or system solution fulfills all assigned tasks and user objectives.	[SY] [UCM] [PFV]
Functional Correctness	The degree to which the software application, product, or system solution produces accurate and consistent results in accordance with the required specifications.	[PRM]

Note:
– These functional qualities can imply suitability, accuracy, interoperability, security, and functionality compliance.

Functional requirements can be made explicit using the *requirement traceability matrix* (between requirements and use cases), page flow view, use case model, and business process view, or can be expressed using functional features from an application, product, or system.

Application Service Usability

Usability is defined as the ease with which a user can learn to operate, prepare inputs to, and interpret outputs from a system or component.[6]

---

[6] IEEE and ISO/IEC 2010, p. 388.

Table 4-7 presents common usability compliance metrics, all of which depend on the user interface (UI) service notation. Although the UI, as part of client-side engineering, is the mother of all user-perceived issues, most issues actually stem from operational qualities such as performance, availability, and security.

Table 4-7. Usability Metrics

Metric	Measurement Description	A-ESA
Accessibility	Accessibility is the extent to which a software product can be used to achieve a specific goal in a particular environment. **Example measurement:** Satisfy the Corporate Device and Browser Support Accessibility Guidelines and design a graphic use interface specification for the visually or hearing impaired.	[UI]
Customization	Customization is the act of altering a product or service to meet the preferences or requirements of an individual or organization. **Example measurement:** Support user customization of the user interface, changing the color of the text, and adding personal images.	
Personalization	Personalization supports displaying the appropriate interface based on the user's permissions. **Example measurement:** When the user permissions change, the menus and buttons that the user can see or operate also change.	
Operability or Simplicity	The degree to which an application or system is easy to operate, control and use appropriately. **Example measurement:** Startup/request latency, memory footprint, build/update duration, and operating complexity ratio.	

(continued)

Table 4-7. (*continued*)

Metric	Measurement Description	A-ESA
Learnability	The extent to which an application or system allows users to learn how to use it effectively and efficiently in an emergency. **Example measurement:** A naive user can navigate to menu items with a maximum of three mouse clicks.	
User Error Protection	It is the extent to which the system protects users from making mistakes. **Example measurement:** All errors provide user-friendly, non-technical messages.	
User Interface Aesthetics, Appropriateness	Aesthetics or attractiveness is a subjective dimension indicating the kind of response a user has toward the IT system. **Example measurement:** The degree to which the user interface is pleasant and satisfying to the user via feedback.	
User Satisfaction	The measure of overall customer satisfaction. **Example measurement:** Customer Satisfaction (CSAT), Net Promoter Score (NPS), etc.	

Note:
- There are various forms of usability measurement criteria, some of which measure overall quality in relation to usability. It is up to the solution architect to determine the measurement criteria. For example, here is a typical four-pillar quality measurement scale for an Android application:
  1. User experience: Pleasant to use
  2. Core value: Fun, useful, or both
  3. Technical quality: Get the most out of premium devices
  4. Privacy and security: Designed for safety

Application Service Adaptability

Table 4-8 shows the *major solution quality* metrics in software engineering. Note that the cause of most application solution issues is improper architectural design of these metric artifacts.

Table 4-8. *Application Service Adaptability Metrics*

Metric	Measurement Description	A-ESA
Computational Complexity	Complexity means how intricate and complicated something is. It measures how difficult a system design is to understand and implement. It can mean how complex the code or logic is or how long it takes to run and how many resources it uses. High coupling makes things more complex. **Example measurement:** Number of interfaces between applications, services, or system devices, or the number of service component modules and module states.	[UI] [AS] [DS] [SC] [PTN]
Cohesion	Cohesion shows what a module, a service, or a component is responsible for doing. Cohesion can't usually be measured automatically. It's usually measured by coupling. **Example measurement:** Single responsibilities or responsibility boundaries.	[SC] [SCR]
Service Dependency or Coupling	This means the dependency between services or service components. High dependency results in high coupling. **Example measurement:** The number of efferent and afferent dependencies.	[SIV]

Application Service Maintainability

The non-functional or cross-functional quality metrics of functional services determine the *application maintainability* of the solution. Table 4-9 lists the major maintainability metrics for application services and their associated artifact notations.

Table 4-9. *Application Service Maintainability Metrics*

Metric	Measurement Description	A-ESA
Reusability	Reusability is how easy and fast you can reuse a portion (or more) of a system design and/or implementation. In other words, it is the extent to which a service can be used in multiple modules, systems, or platforms. **Example measurement:** A service component can be understood and reused by an entry-level developer within 30 minutes, or a service component can be used in more than one application.	[UI] [AS] [DS] [TS] [SI] [SC] [DP] [MS]
Modularity	Modularity means that the system is made up of different parts that can be changed without affecting the others. **Example measurement:** The maximum cyclomatic complexity of a module should not exceed 22.	[DP] [SC]
Verifiability	Verifiability is how well software components or the product can be evaluated to show that it works. It's related to testability. **Example measurement:** The mean time to find a failure, assuming test equipment, facilities, and personnel are available.	[AP] [SI] [DP] [PS] [MS]
Testability	Testability is how easy it is to test an application and how well the tests interact with the code to find bugs. **Example measurement:** The test environment contains realistic seed data and simulates real workloads.	[AP] [SY] [MS]
Analyzability	Analyzability is an assessment of how changes to one or more parts affect other services, modules, and systems. **Example measurement:** The solution is readable (correct interpretation) and understandable (semantic connotation) by a specified stakeholder.	[VLD]

(continued)

Table 4-9. (*continued*)

Metric	Measurement Description	A-ESA
Modifiability or Changeability	Modifiability is how easy it is to understand, change, and expand software designs and code. **Example measurement:** Changes are anticipated before the architectural design, and the extent to which a service or service component can be modified effectively and efficiently without introducing defects or reducing the quality of existing services.	[AP] [SI] [SC] [DP] [MS] [PD]

Note:

– Maintainability can also be categorized as a non-runtime quality, as shown in Table 4-11, which reflects a
different viewpoint.

– Reusability and modularity can also be part of adaptability.

Operational Quality Metrics

The architecture model is not a *systems engineering* work until it is operationalized. Operational views reflect the solution system in its runtime environment. These operational qualities require consideration of both functional and operational artifacts. Operational quality attributes can be divided into *runtime (user-aware)* and *non-runtime* qualities, although the detailed taxonomy varies.

There are many operational quality metrics, especially in cloud platforms. The following section lists some commonly used ones from a *non-functional requirements* perspective.

User-aware Operational Quality

Table 4-10 is a list of major *user-aware* (*runtime* or *operational*) quality attributes and their associated artifacts for the solution architect to ponder. This is the technically challenging area. It needs to be considered along with the overall architectural quality and the application service quality (both application logic and data services). Many of the operational quality issues in the production environment are actually caused by lousy requirements modeling up front.

Table 4-10. *User-aware Operational Quality*

Category	Metric	Measurement Description	A-ESA
Performance	Duration	Duration is the lasting time in the UI, app, memory, CPU, hardware, network, etc. **Example measurements:** – Elapsed time: 30 seconds – CPU time: 2 seconds – Wait time: 1 second	[SY]
	Capacity	Capacity is the extent to which the maximum limits or volumes of the services meet the specified requirements. **Example measurements:** – Capacity planning, stress testing, SLA relevance – Estimation of significant business application process throughput – Prediction of each node's peak volume based on the historical data – The website should be able to handle a page view growth rate of 50% per year without performance degradation. – The distribution solution should be able to accommodate up to five new warehouse centers per year. – Minimum capacity in the UNIX environment: 300 servers and 60 active applications – Use distributed load testing tools to generate peak traffic or estimate the workload at the planned growth rate	[MW] [DB]

(continued)

Table 4-10. (*continued*)

Category	Metric	Measurement Description	A-ESA

		– Volume metrics
		– Storage required: 10TB of Tier-1
		– Data transferred: 3000MB
		– Page Views (PV): 30,000
		– Peak UV (Unique Visitors) per hour: 10 million
		– Maximum concurrent users: 180,000
		– Number of requests per second (RPS): 10,000
		– Number of queries per second (QPS): 25,000
		– Peak QPS calculation:
		  – If total PV= 5,000,000, and 70% visits per day in 30% peak time
		  – Peak QPS = (Total PV * 0.7) / (Seconds a day * 0.3)
		  – Peak QPS = (5,000,000 * 0.7)/ (24*3600 *0.3) = 135
		– Number of hardware devices to calculate:
		  – If a single hardware device, QPS = 80
		  – Number of hardware needed = Peak QPS / single hardware device QPS
		  – Number of hardware calculation: 135/80 = 1.73
		  – Number of hardware needed: 2	

	Response Time	Response time is a measure of how well a service performs its functions within the required time constraints and throughput. **Example measurements:** – Page response time requirement: less than 1 second – Stress test response time: 0.03/s	[DEP]

(*continued*)

Table 4-10. (continued)

Category	Metric	Measurement Description	A-ESA
	Throughput	Throughput is the amount of data processed by a machine or system in a given period of time. The stage with the lowest throughput bandwidth determines the overall tps (transaction per second) throughput. **Example measurements:** – Throughput: 30MB/s for the significant flows such as order processing – Requests per second: 1,000,000 RPS (peak)	[DEP]

	Concurrency	Concurrency is when two or more things happen at the same time. Concurrency is equal to QPS * average response time, and it includes consideration of locks and semaphores. Asynchronous processing and event handling are also concurrency considerations.
		
		Example measurements:
		– Request per second: 1000
		– Container concurrency: 60
		– Transaction:
		  – Transaction rate: 3000/s
		  – Number of transactions: 100,000 hits
		  – Longest transaction: 3.03
		  – Shortest transaction: 0.03
			[DS]

	Latency	Latency is a delay in data transmission or processing caused by network, disk, or CPU issues. It measures the slowest requests. **Example measurement:** – The latency for the service is 500ms at 99.9%.	[NW]

(continued)

Table 4-10. *(continued)*

Category	Metric	Measurement Description	A-ESA
	Efficiency	Efficiency is how much resource the application uses to complete tasks. Efficiency and performance are linked. If a system uses too many resources, users will have poor performance or low optimization. It can be measured using RPS, QPS, and business metrics (maximum x per second, percentiles). It is a measure of how efficiently the system uses processor capacity, disk space, memory, or communication bandwidth. **Example measurements:** – Fast time behavior – Average support ticket resolution time – Percentage of resource utilization, such as CPU utilization, memory utilization, and disk I/O – Efficiency compliances such as workload resource consolidation – Data de-normalization – Right-sizing to match workloads and avoid over-provisioning – 6GB network traffic per transaction – 1 million transactions per minute (tpmC[7]) – Number of resources provisioned for each unit of work = resources provisioned (proxy metrics) / business results (business metrics) – vCPUs (VM) minutes per transaction – The relationship between the performance and the resources consumed – The need to import 200,0000 records in two minutes on two CPUs – Offloading data to caches and offloading processing in the background	[ND] [VS]

(continued)

---

[7] tpmC (transactions per minute) is a benchmark rating from the TPC (Transaction Processing Performance Council).

Table 4-10. (*continued*)

Category	Metric	Measurement Description	A-ESA
Availability	Availability	Availability generally refers to the quality or condition of being able to be used or obtained (operational locations, online timeframe, etc.). This definition can cover a wide range and has many solutions like CDN.[8] There are many possible causes: resource exhaustion, unplanned load-based changes, and increased number of moving parts (application complexity), external dependencies such as SaaS services, infrastructure, or cloud-based services. Availability is the product of the availability of all elements. **Example measurements:** – Site availability percentage = (total seconds in period) – (seconds system is down) / (total seconds in period) – Sum of all requests received and successfully processed from all APIs in the service to calculate availability in one five-minute interval – Availability based on requests = successful responses / valid requests – Automated service quota management for the cloud workload – Graceful degradation from hard dependencies to soft dependencies – Specify timeouts, retry limits, or throttle requests such as API endpoint throttles – If the order times out and is not paid, the cancellation timeout activity is automatically closed – Fulfillment service timeout is 500ms	[SY] [DB] [NW] [ND] [DEP]

(*continued*)

---

[8] A CDN (Content Distribution Network) is a geographically distributed network of proxy servers and their data centers for high availability and performance.

Table 4-10. (*continued*)

Category	Metric	Measurement Description	A-ESA

		– Related availability metrics:
		  – Total availability: 99.99%
		  – System service hours: 7*24
		  – Archived storage time of ten years for order data
		  – Minimum up-time 99.9% during business hours
		– All data on servers should be deleted and made unrecoverable without being sent to the trash
		– Minimum response time of 0.1 milliseconds per each transaction
		– Mean Time Between Failures (MTBF) = MTTF + MTTR
		– Mean time to failure (MTTF) = total lifespan across devices / # of devices: 1.23 years
		– Mean time to repair (MTTR) = total time spent repairing / # of repairs:5 minutes
		– Mean Time To Detect (MTTD) = total time between failure and detection / # of failures
		– Mean Time to Recover/Resolution (MTTR) for Agile architecture	

	Scalability	Scalability is the ability of a system to accommodate an increasing workload, including horizontal scaling (scaling out) and vertical scaling, as well as distributed database scaling. – Related is *application-level load balancing,* which intelligently provides scalability, performance, and availability. – Also related is *network load balancing,* which efficiently distributes incoming network traffic across a group of backend servers.	[SY] [DB] [ND] [MW] [NW] [DM]

(*continued*)

Table 4-10. (*continued*)

Category	Metric	Measurement Description	A-ESA
		Example measurements: – Support sales targets for the next three years – Scale support for identified application, data, and technical domains – Use of resilient solutions in worst-case scenarios	
	Elasticity	Elasticity refers to robustness, fault-tolerance, and resilience. – *Robustness* is the ability of a system or component to function correctly in the presence of invalid inputs or stressful conditions. **Example measurement:** – The system should be able to handle load spikes – *Fault-tolerance* is how well services keep working even when hardware or software fails. **Example measurement:** – An uninterruptible power supply (UPS) should be available to protect the system from primary power failure. – *Resilience* is the ability of a dynamic system to adapt to disturbances. Some resilience controls help detect problems, while others help fix them. Compared to fault-tolerance, resilience responds more gracefully to unknown failure modes. **Example measurement:** – System enables self-healing capability	[SY] [MW] [ND] [MS]

(*continued*)

Table 4-10. (*continued*)

Category	Metric	Measurement Description	A-ESA

Security	Data Integrity	Data integrity is the assurance that data is not lost, duplicated, or mistakenly exchanged when used with other services or systems.
		
		Example measurements:
		– Interface communication, synchronous calls including idempotent operations, transaction integrity, especially during distributed transaction, compensating transaction:
		  – Successful transactions: 100,000
		  – Failed transactions: 0
		– Mismanaged software versioning
		– Unauthorized application updates
			[DS]
			[DB]
			[SIV]

	Authenticity and Authorization	Authentication verifies that users are who they say they are. Authorization checks if users or systems can access resources. **Example measurements:** – Single Sign-On and Identity and Access Management (IAM) – Certificate revocation listing	[DS] [TS] [MW]
	Data Security	Data security includes data security in transit, digital signatures, and encryption. **Example measurements:** – Data confidentiality from external disclosure – Data classification marking – Secret rotations – Storage encryption is required – Solution must comply with EU GDPR standards	[DS] [DEP]

(*continued*)

Table 4-10. (*continued*)

Category	Metric	Measurement Description	A-ESA

	Assess Control	General access control includes five main security goals: confidentiality, availability, authentication, integrity, and non-repudiation. Network access security prevents internal resources from being exposed to external users.
		
		Example measurements:
		– Implement role-based access control
		– Establish proper security segmentations
		– Vulnerability management and regular scanning for known computer viruses, worms, SQL injections, and malicious attacks
		– Deploy security information and event management (SIEM) and workload security tools
		– Multiple security controls at all levels
		– Reduced security attack surface
		– Regular penetration testing
		– Network security solutions:
		  – Intrusion Detection System (IDS)
		  – Intrusion Prevention System (IPS)
		  – Distributed Denial of Service (DDoS)
		  – Web Application Firewall (WAF)
			[SI]
			[NW]

	Non-repudiation	Non-repudiation means that the integrity and origin can be verified and validated by a third party as coming from a specific entity. **Example measurement:** – Multi-Factor Authentication (MFA)	[DS] [UI]

Note:
– Only define metrics that are critical to your solution and set a baseline for each metric with a likely range, such as worst and best case.
– Identify metrics that need to be revisited periodically to adapt to changing situations.

Non-runtime Operational Quality

Table 4-11 is a list of non-runtime quality attributes that are handled in the backend and need to be planned for in advance to ensure smooth online business operations.

Table 4-11. *Non-runtime Quality*

Category	Metric	Measurement Description	A-ESA

Manageability	Configurability	Configurability is changing how application services work by changing the settings.
		
		Example measurements:
		– Ensure that the newly added functions or fixes are configurable
		– Externalized configuration: The business administrator must be able to perform on the business interface and the business management without resorting to technical means such as SQL statements
		– Batch and emergency configuration must be automated
		– Use Infrastructure as Code (IaC) or Operations as Code (OaC) for operational configurations
		– Use a cloud architecture change management:
		  – Google Cloud Asset Inventory
		  – AWS Config
		  – Azure Application Change Analysis
			[SY]

(continued)

Table 4-11. (*continued*)

Category	Metric	Measurement Description	A-ESA
	Traceability	Tracing is a record of application services health information. **Example measurements:** – There is a clear structural mapping between tracing service components – A UI test is able to trace backend systems when a failure occurs – Ensure automatic detection of sub-system errors – Auditability: Audit logs for the core operations (such as logins and order cancellations)	[DEP]
	Recoverability	Recoverability is how well a system can recover data and restore its normal state if it stops working. An effective incident response procedure must be implemented. **Example measurements:** – Time of backup: 23:00 EST – Backup frequency: daily – Recover snapshots: 2 – Retention periods: 60 days – Archiving period: 1 year – Recovery Time Objective (RTO): 24 hours – Recovery Point Objective (RPO): 1 hour – Recovery Point Actual (RPA): 30 minutes	[PTN]

(*continued*)

Table 4-11. *(continued)*

Category	Metric	Measurement Description	A-ESA

	Observability	Observability shows a system's health, internal state, and context. The three pillars of observability include *metrics, logs,* and *traces.* Observability dashboard collects timeline data and track patterns over time. Usually, observability comes with monitoring (data collection). To be effective, a system-wide monitoring strategy needs to be implemented.
		
		Example measurements:
		– Capture software runtime events and historical events and report or notify them by alerts
		– Maintain security observability data for two years
		– Include application performance monitoring (APM)
		– Common metrics such as:
		  – Sum of traces
		  – Mean erroneous traces rate
		  – Mean latency, calls/second
			[SY]
			[MW]

	Deployability	Deployability indicates the ease of deployment in a target environment. **Example measurement:** – Automated deployment and runbook for rollback	[DEV] [DP]
	Portability	Portability measures the effort required to migrate software from one operating environment to another. **Example measurements:** – The effort (human, technical, cost, and environmental factors) required to transport and adapt it to a new environment is less than the effort required to redevelop – Maintain the portability of the application layer program code across operating systems	[ND]

(continued)

Table 4-11. (*continued*)

Category	Metric	Measurement Description	A-ESA
Maintainability	Maintainability	Maintainability is how easy and fast a system can be changed for better, more adaptive, corrective, and preventive reasons. **Example measurements:** — Mean time between planned preventive maintenance — Solution should be modeled and updated using architectural design tools — User interface should be easily customizable	[SY]
	Extensibility	Extensibility is the ability of a system to adapt to future changes. This can be achieved through flexible architecture, design, or implementation. **Example measurement:** — The solution should be able to accept new functionality from future requirements	[SRV] [AOV]
	Documentation	Documentation helps to explain the system. **Example measurements:** — Documentation of architecture decisions — A clear checklist of solution metrics and measures — Model deliverables with key property specifications — Simulated or tested disaster recovery plan — Design guidelines such as naming and code conventions and design patterns	[DL]

(*continued*)

Table 4-11. (*continued*)

Category	Metric	Measurement Description	A-ESA

Reliability	Reliability	Reliability is the extent to which an IT system can operate without failure within a specified timeframe and environment. It refers to the capability of software to maintain its level of performance for a period of time.
		
		Example measurements:
		– High value of Mean Time to Failure (MTTF)
		– Runtime redeployment of a modified service with no negative impact
		– Using Failure Mode and Effect Analysis (FMEA) to prevent potential failures
		– Reliability through geographic replication
		– Dependable factors such as:
		  – Maturity
		  – Fault-tolerance
		  – Recoverability
			[GV]
			[SY]

	Stability	Stability refers to the quality of a system, application, or service to maintain consistent performance and reliability over time without significant issues or disruptions. **Example measurement:** – Hardware or software stability with no runtime hiccups	[AP] [SY] [MW] [ND] [NW]
	Dependability	Dependability is the consistency of the measurement. **Example measurements:** – Runtime redeployment of a modified service with no negative impact – Nodes migrated to the X cloud must retain their current IP addresses	[DEV]

(*continued*)

Table 4-11. (*continued*)

Category	Metric	Measurement Description	A-ESA
	Redundancy	Data redundancy occurs when the same piece of data is stored in two or more different locations. **Example measurements:** — Data replication must be in place for certain data services — Use heavy caching wherever and whenever possible — Significant business flows must be redundant — Production databases must be highly available with a passive node at site X in a secondary availability zone — Ensure redundant connections between on-premise platforms and private networks in the cloud	[DP] [DB] [MW] [ND] [NW] [VS] [CA] [ZN] [MS]

Note:
— There are a huge number of metrics. Metrics for specific service-level characteristics are beyond the scope of architectural modeling. The same is true for granular metrics, such as replication metrics in the cloud environment: operations pending replication, replication latency, etc.
— Some non-runtime quality attributes are closely related to runtime quality attributes, such as non-runtime redundancy to runtime availability.
— The categorization can vary from solution to solution. For example, security may be further subdivided into runtime and non-runtime (data security at rest or static security testing) or a cross-cutting category.

DevOps Metrics

DevOps, as a modern enterprise solution environment, is naturally part of the measurement consideration. The DevOps environment is a critical place to test the non-functional hotspot issues.

There are many Agile metrics (such as DORA for DevOps), including product lead time, team velocity, issue cycle time, satisfactory score (net promoter score), iteration timeline (days), epic and release burn-down and burn-up (charts), cumulative flow diagram (CFD, the state of the work in progress, work item age, throughput, blocked time, time box iteration, escaped defects, percent of CI build failures, number of times code is checked in, automated code review errors, failed deployment, code coverage, test automation rate, team efficiency (velocity/number of person days worked), pair programming, and so on.

Table 4-12 demonstrates how A-ESA ensures architectural quality in the DevOps environment, where test or monitoring tool *middleware* (MW) is used to keep track of common solution requirement (RQ-NFR) metrics. The *key consideration* (KC) element is the action element that specifies how to resolve risk (RK) issues, from architecture, design, or implementation.

Table 4-12. *DevOps Metrics Using Middleware Tooling*

NFR	RQ Metric	MW Tools	RK (Hotspot)	KC Element
Availability	Uptime	HyperSpin	TBD	TBD
Reliability	Error Rate	Splunk	TBD	TBD
Performance	TTLB (Time to Last Byte)	Fiddler	TBD	TBD
Maintainability	Code Coverage	SonarQube	TBD	TBD
	Complexity	SonarQube	TBD	TBD
Security	Static Application Security Issues	CheckMarx	TBD	TBD

Note:

– TBD: To be determined after testing.

– A hotspot (RK) element is any functional or operational element that causes a major issue. For example, in the TTLB (HTTP throughput performance) test, a hotspot element can be a UI or/and a NW element.

– Maintainability Index (0 - 100) is sometimes used to represent the relative ease of maintaining the code.

In addition to measuring the system solution, DevOps also requires the measurement of solution development objectives and key results (OKRs). Table 4-13 shows an example comparison between the traditional KPIs and DevOps OKRs. While KPI is an intent-oriented requirement, OKR (action item) is more of a results-oriented governance approach. OKRs can be grouped into categories: system performance and resilience, work practices, developer support, incident response, DevSecOps, FinOps (cloud cost control), work practices, and so on.

Table 4-13. *Comparison of KPIs and DevOps's OKRs*

Metric	Requirement Goal	Measurement Result	A-ESA
KPI	Complete 100 ticket responses per hour.	Completed a response rate of 100 tickets, but increasing volume resulted in slow response times	[RQ]
OKRs	Objective: Solve the support ticket crisis		[GV]
	Key result 1: Reduce response times by 10%	Reduced response times by 8%	
	Key result 2: Improve customer satisfaction score by 10%	Improved customer satisfaction score by 6%	
	Key result 3: Increase the number of tickets handled from 90 to 100	Increased the handled tickets from 90 to 96	

Note:
- Assume a quarterly review for both KPIs and OKRs.
- In the case of the KPI, the performance index is static, resulting in longer customer wait times and lower customer satisfaction.
- In the case of OKRs, while the key results have not been met, the dynamic nature of the OKRs has significantly improved the support ticket crisis.
- The difference between a KPI and a metric can be trivial, but typically KPIs are a select set of metrics designed to meet business objectives.
- OKR frameworks can be part of governance.

Often, architectural issues in the DevOps environment include:

- A team member may wear multiple hats, so try to avoid eating your own dog food and let outsiders to do the reviewing.

- Developers generally don't have a strong sense of long-term quality metrics, and they aren't rewarded for it, so DevOps has to go beyond its scope for a viable architecture.

- If DevOps with constant releases frequently changes the modeling landscape, then check how this affects the stability of the system.

DevOps is where software engineering meets systems engineering. It requires both hard and soft skills. In addition to following the automated validation process and meeting productivity criteria, a DevOps solution architect needs to create simple model mappings and define guardrails to ensure architectural conformance in a results-driven environment.

SRE Practice

SRE (Site Reliability Engineering), a discipline developed within Google, is a more specific way of doing DevOps. In an Agile IT with digital systems, IT software development, IT support, and IT infrastructure have become very blurred, be it DevOps, CI/CD, or SRE. In recent years, AIOps and ChatOps have become part of it.

Key SRE KPIs include service level indicators (SLIs), service level objectives (SLOs), and defect budgets.[9] Engineering OKRs and DR metrics (MTTD, MTTR, and MTBF), as mentioned earlier, are also considered. The key preventive approach to SRE is chaos engineering, guided by model-significant case walkthroughs or, a step further, by model validation.

---

[9] It's how long a system can be offline without breaking its SLAs or SLOs.

With all the tools, technologies, and automation within SRE, you might think that operational architecture is a breeze. Yes, SRE takes a lot of the pain and hassle out of the traditional development and deployment process. But consider the following points:

- The impact of SRE depends on the culture of the organization, including breaking down silos, embracing risk, and improving continuously.

- AIOps is about data analytics and machine learning. The usefulness of AIOps depends on the required data and architectural guardrails.

- Even if a well constructed SRE ensures the reliability of the runtime platform, many functional or logical issues still pose significant challenges.

Although SRE design is not the focus of A-ESA, ESA architectural choices (architectural styles and patterns, metrics, division of labor, and so on) impact SRE effectiveness through governance, application development, deployment mapping, software service selection, and Infrastructure as Code implementation.

In a cloud environment, DevOps or SRE is inherently part of the A-ESA views. In a self-managed environment, an A-ESA profile on SRE should be created for clearer architectural considerations.

Solution Constraint and Opposing Force

In an enterprise solution environment, an NFR is *rarely* considered on its own. A common pitfall when solution modeling is to focus on technical measurements rather than *constraints*. This section briefly describes solution constraints and the tradeoffs between requirements and countervailing forces (especially business constraints).

Solution Constraint Metrics

Constraints, or potential risks, are imposed by the solution project
environment, enterprise environment, or the ecosystem environment.
They are the counterforce to the requirements, although the constraints
could be categorized as a general NFR that is part of the requirements.
Table 4-14 gives some examples of three major constraint categories, of
which economy is the biggest constraint in terms of cost.

Table 4-14. Example Constraints and Metrics

Category	Example Statement	Example Item	Example Metric
Economy	The solution must be completed within eight months	– Budget (vs. cost) – Scope – Resource – Risk factors	– Affordability – Profitability – Timeliness – Resource utilization – Productivity
Business	The solution must conform to EU GDPR standards	– Organizational culture – Legal and compliance – Localization – Marketplace factor	– Time to market – Compliance – Localization – Sensitivity – Legality – Flexibility
IT	The application must use Microsoft .NET Framework 4.5. (architecture constraint)	– Development skills – System Environment – Implementation standard – Innovation-mindedness	– Realizability – Maturity – Observability – Safety – Standardizability

Note:

– There is no absolute demarcation between constraints and quality. For example, a constraint such as compliance could be described as a security requirement.

234

Cost as a First-Class Metric

As mentioned, budget is a major business constraint. Cost is the other side of the budget coin. Cost is a factor by which all quality requirements are constrained. Human effort, resource increases, scope creep, and time shifts are all part of cost considerations. More often, cost is used to indicate what is economically viable and justifiable. For solution architecture, cost can be treated as a *first-class* metric. Table 4-15 shows how cost considerations are reflected in the solution architecture.

Table 4-15. *Cost Considerations in ESA*

Cost Consideration	A-ESA
Specify requirement metrics including the cost of services, utilization, security, and performance availability.	[RQ]
Identify cost from different views: users, development teams, solution features, and operational environment.	[UCM], [DEV], [SRV], [DEP]
Determine key cost decisions (working together with solution manager): unit cost, idle cost, cost sharing/efficiency, innovation/cost ratio, time to market, cost spike, etc.	[KC], [MTS]
Decide on a CI/CD approach.	[DEV]
Right-size computing resources.	[ND]
Track and monitor solution (either in-house or cloud environment).	[MW], [CL]
Note:	
– For cost analysis simulations, plug cost estimates or statistics into the property attributes of A-ESA's views and elements, if available. With increased automation and AI, cost intelligence will be part of solution. – When using multiple cloud providers, use a single source of truth for the most critical information.	

Obviously, cost is an enterprise's top concern, beyond the mere enterprise solution consideration. From a modeling perspective, cost is driven by the value proposition. See Table 6-34 in Chapter 6 for cost control and management.

The cost consideration is further elaborated as a key measurement in the Key NFRs (see Table 4-18) and in tradeoff analysis (Table 6-11 in Chapter 6). Note that depending on value propositions, solution architectural styles, and hosting models, cost considerations will differ. ESA modeling focuses on solution architecture, but remains closely aligned with solution management.

Opposing Force Metrics

Constraint covers both *business* and *technical* spaces. It's the counterforce to the requirement. Another requirement can also become a counterforce. Table 4-16 illustrates an architectural consideration of a counterforce balancing scenario for a Double-11 sales day event, which is a tradeoff case between NFR and constraints.

Table 4-16. *Opposing Force Tradeoffs for the Double-11 Sales Day*

Metric	Measurement Description	A-ESA
Scalability Requirement	The system must support 20,000 new users.	[RQ]
Business Constraint (Opposing Force)	The system should be launched before the 2023 Double-11 online shopping day.	[RK]
Technical Constraint (Opposing Force)	The system must leverage the existing legacy system for order management.	[RK]
Architecture Consideration	Assumes there are established middleware capabilities and DevOp skills.	[KC]
	Alternatives: 1) Scale down functional features and implement resilience capability. 2) Re-architect and migrate the legacy application.	
	Decision: 1	

Note:
– The cost constraint is the major countervailing force in balancing quality requirements.

Lean Measurement Cluster

The previous measurement criteria can be referenced as a detailed checklist to determine the significant requirements of the solution. Often, it is more effective for solution management to use only a few categorized terms.

S-MAPS Measurement Cluster

The non-functional requirements listed in Tables 4-10 and 4-11 can be condensed into four categories: performance, security, availability (including non-runtime reliability), and manageability (including non-runtime maintainability). These quality categories plus the sustainability consideration form a simple measurement cluster, as described in Table 4-17. For ease of recall, they can be abbreviated as an acronym: *S-MAPS* (meant by *s+maps*), where *sustainability* is a long-term architectural goal, and *manageability*, *availability*, *performance*, and *security* come from the operational quality requirements. All A-ESA notation mappings need to follow S-MAPS.

Table 4-17. *S-MAPS NFR Description*

S-MAPS	Description
Sustainability [RQ-SUS]	Sustainability encompasses a broader range of factors (environmental ecosystems, green IT architecture, or fair distribution of resources and benefits), so it often requires consideration of both cost effectiveness and cost efficiency, or in simple terms, *cost optimization*. For A-ESA, a sustainable architecture ensures economic viability and resilience, as well as long-term financial stability and growth.
Availability [RQ-AVL]	Availability is closely related to *reliability*, which is a non-runtime quality, and to scalability and resilience. Both reliability and scalability have an impact on availability. Therefore, they are included here as part of the availability considerations.
Performance [RQ-PER]	Performance relates to capacity, efficiency, concurrency, latency, throughput, etc.
Security [RQ-SEC]	Security covers authenticity and authorization, access and transmission protection, non-repudiation, data reliability, etc.
Manageability [RQ-MGT]	Manageability has both runtime and non-runtime attributes. For example, monitoring is a runtime quality, and configuration is a non-runtime consideration. Close to manageability is another M—*maintainability*—which is a non-runtime quality consideration. In comparison, manageability refers to the operation of a system, while maintainability refers to changes to the system itself. For a loose definition, I include maintainability here as part of the manageability category.

Note:

– Many well architected frameworks use reliability as one of the pillars of quality. Availability, however, is user-centric, taking into account not only the reliability of the system, but also its ability to recover. Therefore, end users are more likely to perceive availability than system end reliability.

– This simple categorization is a common practice. It also resembles many cloud well-architected frameworks. When adopting a cloud-hosted solution, cost is commonly a major decision.

Note that while sustainability is a relatively an abstract term, cost effectiveness, cost efficiency, and cost optimization are easy to understand. In a broad sense, sustainability is about achieving better architectural results within budget constraints.

S-MAPS NFRs Checklist

This section provides a list of examples of architectural considerations and measures of S-MAPS NFRs mapped to A-ESA notations. See Table 4-18.

Table 4-18. *S-MAPS Architectural Considerations*

S-MAPS	Architectural Considerations	A-ESA
Sustaining Cost Optimization	Cost efficiency	[PR]
	Budget limits	[RK-CST]
	Cost responsibilities	[RO]
	Economy of scale or economy of scope	[VL]
	Architectural styles and building blocks	[KC]
	Operational component cost	[DEP]
	Functional component cost	[SRV]
	Development and deployment cost	[DEV]
	Objectives and priorities	[DL]

(*continued*)

Table 4-18. (*continued*)

S-MAPS	Architectural Considerations	A-ESA
Manageability	Operational automation	[MW]
	DevOps culture	[ORG]
	Infrastructure as code (IaC)	[ND]
	Maintainability	[GV]
	Operational responsibilities	[RO]
	Configuration and operational procedures	[GV]
	Simulation and QA processes	[GV]
	Helpdesk support procedures	[GV]
	Emergency and disaster recovery processes	[PS]
	Health observability metrics	[RQ-NFR]
Availability	Stability and integrity	[RQ-NFR]
	Scalability and resiliency	[RQ-NFR]
	Prevention of single point of failure for significant cases	[DEP]
	Functional service availability and reliability	[SIV]
	Data transactional integrity	[PTN]
	Integration and disintegration strategy	[PTN]
	Self-healing capability	[MW]
	Independently deployable services	[MS]

(*continued*)

Table 4-18. (*continued*)

S-MAPS	Architectural Considerations	A-ESA
Performance	Predicted changes for capacity planning	[GV]
	Performance optimization	[GV]
	Performance metrics	[RQ-PER]
	Identification of performance hotspots	[DEP]
	Service component optimizations	[SCR]
	Optimized performance for significant business functions	[UI]
	Real-time performance optimization	[GV]
	Caching management	[MW]
	On-demand degrading and scaling	[GV]
Security	Security baselines and guardrails	[GV]
	Prevention of viruses, access attacks, etc.	[MW]
	Secured development lifecycle	[DEV]
	Protection of sensitive information	[DS]
	Security monitoring	[MW]
	Security incident response procedures	[GV]

Note:

– This is an initial overview of solution NFR considerations. You can drill down for more detail using a different approach, such as Information Security Practice (Table 5-49 in Chapter 5) for security, or large website architecture (Tables 5-36 and 5-37 in Chapter 5) for high availability and performance. All of these five NFRs affect each other, and the tradeoffs between them can be seen in Table 6-11 in Chapter 6.

S-MAPS NFRs by Team Responsibility

IT solution architecture measurement is all about NFRs. There is no right or wrong judgment for categorization, and the key is to connect the dots of all key solution NFR concerns. Table 4-19 shows an initial solution model mapping using core NFR clusters (S-MAPS) through minimal overall categorization by team responsibility.

Table 4-19. *S-MAPS NFRs by Area of Responsibility*

Category	Core NFR	A-ESA Notation
Constraint	Sustaining Cost Optimization	– Value [VL] (value stream analysis) – Risk/Constraint [RK] (counterforce analysis) – Key Considerations [KC] (long-term cost/benefit)
Runtime	Performance Availability Security	– Functional Service View [SRV] (significant case what-if analysis) – Operational Service View [DEP] (hotspot analysis) – Key Choice [KC] (tradeoff analysis)
Non-runtime	Manageability	– Governance [GV] (governance analysis) – DevOps [DEV] (CI/CD pipeline readiness)
Note: – Constraints can also come from Capability [CAP] (heat-map analysis), Role [RO] (responsibility, skill, and culture analysis), etc.		

Teams can identify their concerns in their responsible category, likely using various assessment frameworks as a checklist. For example:

- Cost estimation can be closely observed through workload usage in the cloud platform.

- Runtime quality attribute analysis can reveal patterns, benchmark and simulation walkthroughs, decision records, and so on.

  – What-if analysis for functional services uses a simulated failure approach similar to chaos engineering.

- – Operational service hotspot analysis identifies potential bottlenecks or risk points.

- Non-runtime analysis can result in service component realization guidance, and so on.

- All analyses can be correlated as architectural strategies via optimized metrics views.

Summary

This chapter has presented a list of solution NFRs. For a particular solution, you may focus on only a few NFRs, but knowledge of all the major NFRs, their mapping to the solution architecture, and their cross-cutting implications is a must for every ESA architect.

Note that many architectural tradeoff decisions are made on interdependent NFRs, and architectural debt is often incurred by missing or neglecting related NFR considerations. For example, stability means consistent performance, reliability, fault-tolerance, predictability, scalability, resilience, security, low error rates, and so on. In addition, emergent behaviors may arise from the interactions of different components within the solution system rather than being expected. Therefore, managing NFRs involves architectural strategies, careful consideration of overall quality attributes, proper selection of architectural styles, dynamic model mapping, and continuous feedback and improvement to minimize potential risk and ensure system quality.

Note that requirements are also part of solution management. See requirements management in Chapter 6 for requirements management techniques, prioritization, and so on.

The next chapter discusses architectural styles, each of which has a unique set of effects on NFRs.

CHAPTER 5

A-ESA Modeling Styles

The best architectural style fits the purpose.

There are many architectural styles to choose from. Choosing the right one is critical for enterprise solution architecture and enterprise software systems. In fact, *an architectural style can have a significant impact on solution quality attributes.* Reference architectures, patterns, and best practices highlight some of the architectural style considerations. Table 5-1 provides a list of IT architectural styles.[1]

---

[1] These architectural styles leverage techniques of base architectures (BA, AA, DA, and TA) when needed.

© Sean (Chunhong) Gu 2024
S. Gu, *Mastering Enterprise Solution Modeling,*
https://doi.org/10.1007/979-8-8688-0992-7_5

Table 5-1. *A General List of IT Architectural Styles*

Arch Style	Description
Mainframe architecture	The mainframe architecture is made up of hardware and software parts that work together to process and store data. This architecture is a set of defined terms and rules that are used as instructions to build products. This old architecture is finding new life in the AI era.
UNIX architecture	The UNIX architecture is a computer operating system architecture that reflects the UNIX philosophy. Ken Thompson founded the UNIX philosophy, which is a set of cultural norms and philosophical approaches to minimalist, modular software development.
Client/server architecture	In the client-server architecture model, the server is designed to operate as a centralized system serving many clients. Load balancing is the methodical and efficient distribution of network or application traffic across multiple servers in a server farm. It comes in many forms, including terminal-based environments and rich-client architectures.
P2P application architecture	In a peer-to-peer (P2P) network, two or more computers (peers) pool their resources and communicate in a decentralized system. Peers are equal or equipotent nodes in a non-hierarchical network. This is also called a symmetrically distributed system architecture.
Microkernel architecture	The microkernel architecture is sometimes called a plug-in architecture because it is built on a small core framework that can be expanded with capabilities and functionality. It offers a great way to customize apps and can also help with managing the software lifecycle.
Hexagonal architecture	The hexagonal architecture, also called the ports and adapters architecture, is a software design pattern that's been developed to make it easier to build loosely coupled application components. The goal is to create loosely coupled application components that can be easily connected to their software environment through ports and adapters.
Web application architecture	The architecture of web applications defines the interactions between browser-based applications, middleware, and databases, ensuring that multiple applications can work together.

(continued)

Table 5-1. (*continued*)

Arch Style	Description
Event architecture	The event architecture is a distributed architecture based on decoupled systems that is designed to respond to events triggered and communicated between decoupled services. This architecture has been around for a while, but it's becoming more relevant in the cloud now that serverless architecture is taking off. It can be an event-driven architecture (EDA) or an event-sourced architecture.
ERP architecture	The ERP solution architecture is built on a solid framework and is made up of reusable modules and components for a set of business processes or applications, such as supply chain and manufacturing. It includes the software and the hardware, and one product can be accessed in different ways.
Multitier architecture	The multitier architecture (also called n-tier architecture) is a way of organizing a system so that the presentation, application processing, and data management functions are physically separated. Three-tier architecture is the most common type of multitier architecture.
Monolithic architecture	The monolithic architecture is usually a three-tier application architecture, built as one big unit that can stand on its own and doesn't depend on other applications, or it can be a single, large computer network with a codebase that ties all the business concerns together.
Pipeline architecture	The pipeline architecture is a type of computer architecture in which processing units are divided into a series of stages. Each stage is responsible for performing a specific task and passing its output to the next stage. Although it is called an architecture, it is virtually an architectural pattern. It is commonly used in the ETL (Extract, Transform, Load) application process.
Mobile App architecture	In a simple definition, the mobile app architecture is a combination of model/design and techniques used to build a mobile application ecosystem. It acts as a blueprint for a mobile application that will take shape according to the architecture.
Cloud-based architecture	The cloud-based architecture, part of the hybrid architecture, makes use of cloud features like resource elasticity, a software-defined network environment, high availability, scalability, and auto-provisioning. This style includes SaaS, PaaS, IaaS, FaaS, and serverless architectures.

(*continued*)

Table 5-1. (*continued*)

Arch Style	Description
Service-oriented architecture (SOA)	SOA makes it easy to reuse and integrate software components through service interfaces. Coarse-grained services use common interface standards and an architectural pattern. This architectural style helps cut down on duplicate functionality and improve interoperability with existing functionality. It replaced old-fashioned point-to-point integration with business process management and service integration architecture.
Microservices architecture (MSA)	The microservices architecture has a lot in common with service-oriented architecture. However, what sets microservices apart is that they are made up of small, self-contained services that follow the UNIX approach to modularity and flexibility. These microservices are designed to model a business domain with a self-contained and single business capability. This architecture works better in an Agile and cloud-native environment, especially when it comes to horizontal scalability.
Service-based architecture	The service-based architecture (SBA) is a loose term that generally refers to SOA, MSA, EDM, or EDA in a service-oriented approach.
Modular monolith architecture	The modular monolith architectural style is a middle ground between traditional monolithic and microservices architectures. Modules can be swapped out and reused, and they have a clear programming interface. Focusing on business domains and stacking modules vertically makes code easier to organize and maintain. Spring Modulith splits an application into microservices, dividing it into cohesive, independent modules within a single deployable unit.
Cell-based architecture	The concept of cell-based architecture is based on the notion of organizing services and data into groups, with each group linked to a specific gateway and a set of associated components. This approach is analogous to that of a microservice API collection. In essence, cell-based architecture is a microservice of microservices, or a coarse-grained service architecture.

Table 5-1. (*continued*)

Arch Style	Description
Supergraph architecture	The supergraph is a graph of graphs, which combines multiple subgraphs that are individual GraphQL APIs. These APIs serve specific data domains or functionality. The supergraph architecture permits the construction and scaling of multiple data domains as a single graph of composable entities and operations. This approach combines the best of a centralized system and a federated microservices model.
Big data architecture	The architecture of big data is comprised of four principal layers: data ingestion, data processing, data storage, and data visualization.
Clean architecture	The clean architecture is a system architecture guideline that Robert Martin came up with to help build scalable and maintainable software. It is made up of four layers, from the inside out: entities, use cases, interface adapters, frameworks, and drivers. The inner circle doesn't know about the outer circle.
Shared-nothing architecture	In distributed computing, a shared-nothing architecture is a way of connecting different nodes through a network, with each node being independent. Each data node has a bunch of small partitions that can be moved around.
Mesh architecture	Various forms of mesh architectures such as: – MASA is a mesh architecture of applications, APIs and services. – Data mesh architecture based on four fundamental principles: domain ownership, data as a product, self-service data infrastructure platform, and federated governance. – Network topology where infrastructure nodes are connected to as many other nodes as possible, to make cooperation easier.

Note:
– There are many other architectural styles based on different approaches, technologies, and applicable environments. For example, object or component-based architecture, resource-based/restful architecture, publish-subscribe architecture, blockchain architecture, IoT (Internet of Things)/edge-cloud architecture, Agile architecture, API-centric architecture, attribute-based architecture, rule-based architecture, space-based architecture, etc.

Most of the architectural styles listed in Table 5-1 fit A-ESA. Chapter 2 provided examples of solution profiles using typical IT solution styles for a broader view and flavor. Note that a solution architecture can use multiple architectural styles simultaneously. This chapter summarizes the use and model mapping of the architectural styles in Chapter 2 and beyond.

The choice of architectural style depends on the usage scenario and purpose. The solution architecture needs to accommodate business growing size and complexity. A big bang approach is doomed to failure. If you can't build a monolithic architecture, you can hardly build a large-scale microservice architecture.

There are similarities between different architectures, such as between service-oriented architecture and component-based architecture, between service-oriented architecture and microservice architecture. In general, there is virtually no such thing as a strictly monolithic architecture or microservice architecture in a large solution. You can use *modular monolith* if necessary. Typically, more than one architectural style, or a custom architecture somewhere in the architectural spectrum, is used in a single *enterprise-level* solution architecture.

In fact, there is nothing new under the sun. All architectural styles have notations in common, albeit with different composition and granularity in terms of component, application, service, system, cell, etc. As the saying goes, *division and unity are the nature of the world*, which means that things are constantly changing. Despite the architectural variations, the essence of ESA remains.

So, don't get bombarded with different choices of architectural styles, because new ones are coming out all the time. Since IT tends to follow trends, it should be careful not to choose trendy but inappropriate architectural styles. Each architectural style has its unique applicable environment, and each distributed system is susceptible to the effects of the CAP theorem. There is no silver bullet or right architecture for every generic case. Neither *accidental architecture* nor overcomplicated architecture will achieve the architectural goal: the cost of change. Without deliberate architectural style consideration, the result in most cases will be a largely spaghetti-like *big ball of mud* or distributed monolith.

This chapter focuses on the core and common A-ESA architecture, as shown in Figure 5-1. All ESA should follow through these core styles to form a holistic solution model, although each solution model may take a different style, form, or shape.

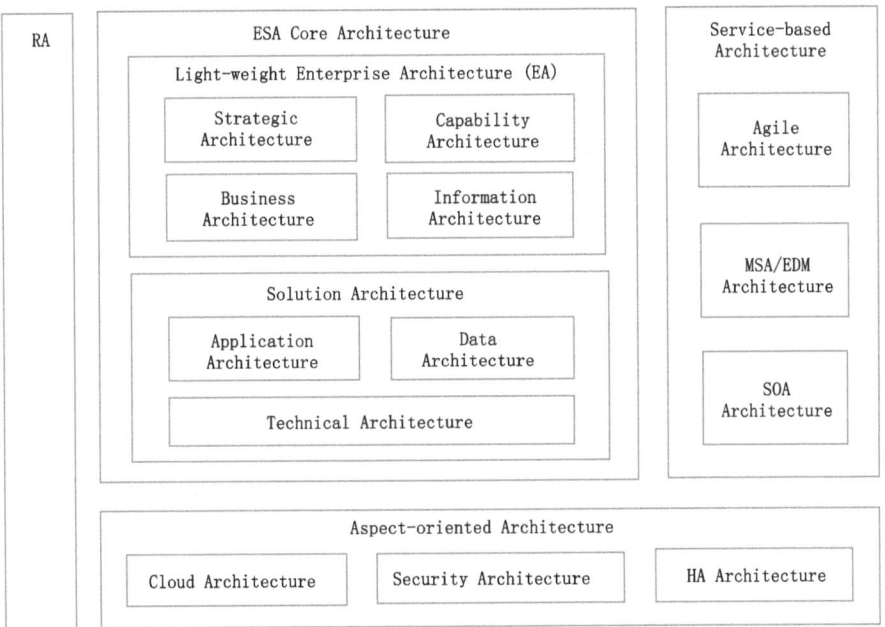

Figure 5-1. *A-ESA architectural styles*

Note that this chapter does not provide a detailed description of each style, but rather focuses on mapping with A-ESA. The mappings of style elements in this chapter are mostly based on real-world cases. They are in *summary* form and require some background knowledge or more in-depth thinking. Each of these styles can be the subject of a lengthy modeling discussion.

HINT

Alignment Between EA and ESA: The architectural mapping presented in this chapter also serves as a litmus test or touchstone for the EA-level input architecture, such as a business architecture. If these high-level architectures cannot be aligned or mapped with the solution-level architecture, they are inconsequential, less useful, or not automatable.

Style 1: Strategic Architecture

ESA blurs EA and functions as a concrete EA in an Agile environment for building an iterative EA building blocks and capabilities. It indirectly considers three environments: *macro-environment* (economy, technology trend, social, etc.), *micro-environment* (competitors, customers, etc.), and *internal environment* (resources, business cases, technologies, and processes). These enterprise environmental factors are reflected in the *IT strategic plan* (ITSP[2]), which plays an important role in the form of IT principles, value streams, or capabilities (or building blocks) in A-ESA modeling.

The strategic *principle* element, which can come from standards and guidelines, is often neglected in a solution environment. The strategic theme is reflected in business *capabilities* (key competitive advantages, innovative ventures, long-term investments) and business *constraints* (pain points, issues, challenges, risks, etc.). The strategic *value* element is tied to cost optimization, which is driven by the value proposition that reflects IT strategic insights: needs, goals, solution offerings and benefits, and differentiators.

Strategic architecture mapping can be partially seen in Solution Profile 1 and is related to the following sections on capability modeling architecture and business architecture. The strategic elements of the ESA are aligned with the IT strategic plan. Strategic architecture elements are intertwined with various architecture views and elements. They are intended for tight solution guidance mappings. In short, the enterprise solution architecture *is guided by* and *influences* the strategic architecture.

---

[2] ITSP includes the long-term strategic directions, mid-term strategic plans, and short-term operational plans.

HINT

Where Is Strategic Direction From? Strategic directions for ESA are succinctly abstracted from enterprise's mission, plans, policies, rules, strategic visions, initiatives, tactics, objectives, goals grid (achieve, preserve, avoid, eliminate), priorities, KPIs/KAIs, value propositions from CxO's speeches and white papers, industry trends, IT trends, competitive analyses (PEST,[3] SWOT,[4] etc.), operating models,[5] market factors, assessments, regulatory compliance requirements, business and technical challenges and issues, roadmaps, and others.

Note that the strategic architecture in A-ESA is narrowly defined to map relevant IT strategic content to architectural elements. It's a lean IT strategic architecture that serves as a set of input elements to the ESA. In fact, strategic architecture can be part of enterprise architecture (EA), and the EA provides a broader context and background for the strategic architecture. The strategic elements are elaborated, detailed, or justified in various forms of EA-level architectures such as capability architecture, business architecture, and information architecture.

---

[3] PEST model (Political, Economic, Social, Technological)

[4] SWOT analysis (Strengths, Weaknesses, Opportunities, Threats)

[5] An operating model is also referred to as an operations model. Don't confuse it with the operational model in a system engineering context.

Style 2: Enterprise Architecture

Enterprise is part of ESA, which is *enterprise architecture (EA)* plus *solution architecture (SA)*. ESA leans more toward solution architecture. It leverages enterprise building blocks if so available from the IT strategic plan or enterprise architecture. As a transition between enterprise architecture and solution architecture, ESA blurs *architecture building block (ABB)* and *solution building block (SBB)* and expresses enterprise architecture content through capability model, reference architecture (RA), architecture overview, metrics view, and so on. Thus, in ESA, EA is related to various architecture views and get grounded in practical mappings.

Table 5-2 shows the major differences between EA and SA. Since ESA is a transition between EA and SA, it covers both ends of EA and SA. Table 5-3 shows example EA mappings in A-ESA.

Table 5-2.   *Illustrative Differences between EA and SA*

Category	Comparison Item	Enterprise Arch	Solution Arch
Target	Objective	Goal [RQ-GL]	Value [VL], Requirement [RQ]
	Purpose	Planning and Assessment [ORG], [AOV], [CAP]	Solution Landing [SRV], [DEV], [DEP]
	Orientation	Intent [AOV]	Result [VLD]
	Deliverable	IT Blueprint [AOV]	SA Model [ML]

(continued)

Table 5-2. (*continued*)

Category	Comparison Item	Enterprise Arch	Solution Arch
Content	Building Blocks	ABB [CP]	SBB [GS]
	Architecture Design	Reference Architecture [PTN-RA]	Pattern [PTN]
	Granularity	Capability [CP]	IT Service [AP], [GS]
	Layering	Grouping [GP]	Cross-cutting [DM]
	Guidance	Principle [PR]	Governance [GV]
	Analysis	Heat-maps [VL]	Significant Case [PS], [UC]
	Concern	Gaps [VF]	NFRs [RQ], [RK]
	Correlation	Association [AN]	Realization [RN]

Note:

− Enterprise Architecture (EA) is generally for enterprise and business people, while Enterprise Solution Architecture (ESA) is for IT architects with EA mapping considerations.

Table 5-3. *Example EA Mappings in ESA*

EA Content	Example Description	A-ESA
Architecture building block	Application architecture (AA), business architecture (BA), data architecture (DA), technology architecture (TA)	[AOV]
	Business capability	[CAP]
	User group	[ORG]
Reference architecture (RA) and pattern	Internet of Thing (IoT) RA	[PTN]
Enterprise guidance	Principle	[MTS]

Experience shows that EA means different things in real enterprise environments, for better or worse. It can be a blueprint, an IT strategic plan, an assessment, a high-level view of IT building blocks, a down-to-earth model, and so on. Or it can be fluffy documentation, a pedantic model, or *shelfware* that is of little or no use. Many organizations honor IT strategic planning and enterprise architecture on paper but not in practice. However, there are other cases where EA adopts common ground standardization, shared capabilities, industry-specific architectural patterns, operational efficiency, and reduced *total cost of ownership* (TCO) through an enterprise-wide IT solution mapping approach.

Style 3: Heat-Mapping Capability Architecture

A-ESA capabilities are not only enterprise or business capabilities but also IT service capabilities. Capability modeling helps to think around the platform (developer platform, business platform, data platform, middle platform, etc.). Although the organizational structure can be roughly aligned with the capabilities, this is not always the case.

HINT

Autonomous Business Capability: In an Agile enterprise, autonomous business capabilities may have their own IT systems and communicate through services, rather than sharing the same IT landscape. This approach, advocated by Roger Sessions, may introduce some inefficiencies, but it greatly improves the speed of change.

Enterprise Capability

Enterprise capability typically includes compressed information about the organization, the value chain,[6] the processes, the value streams,[7] the resources, the course of action, and so on.

The capability is relatively long-lived, while its elements (people, processes, information, and resources) can change frequently. A detailed capability up front may be hard and unrealistic for Agile enterprises. In reality, the ESA capability is formed and adjusted through an iterative and dynamic process to synchronize with the full-fledged enterprise capability model.

---

[6] Value chain is based on cost-analysis and efficiency.

[7] Value streams refer to a set of activities that adds value.

Simply put, a capability model contains a group of capabilities organized by *category* (such as customers, channels, and logistics) or optionally intersected by *level* (such as direct, control, and execute). A capability model can serve several purposes. Table 5-4 shows typical purposes.

Table 5-4. *Purposes of a Capability Model*

Purpose	Brief Description	A-ESA Mapping
Enterprise Strategic Capabilities	It is part of the enterprise roadmaps and blueprints for over years to identify organizational initiatives and execution strategies.	As input mapping to principle [PR] and capability view [CAP]
Enterprise Business Capabilities	It is the business-level capability framework developed by business leaders and domain experts. It embodies an enterprise's business activities, processes or even collaborations. It can be based on industry capability models.	As selective input mapping to capability view [CAP], process [PS], and use case model [UCM]
Foundational Capabilities	It is the capability that applies to all capability models such as stakeholders, governance, techniques, etc.	As input mapping to capability view [CAP]
Enterprise Portfolio Capabilities	It supports cross-solution, cross-projects, multi-phase initiatives which provide an overall landscape of integrated enterprise capabilities.	As an intent-oriented capability model focusing on [GS] and [CAP]
Solution Capabilities	It is a project, product solution-level capability model.	Mapping the high-level capability models such as [AP] and [SY] by segment capabilities

Note:
– Capability can be either conceptual or concrete.

A-ESA is more concerned with IT capabilities that map to business capabilities. Table 5-5 shows two required mapping activities between the enterprise business architecture (EBA) and the enterprise solution architecture. Each ESA capability element may have a set of property specifications (see Table 5-6).

Table 5-5. *Mapping Activities Between EBA and ESA*

Mapping Activity	Measurement Description	A-ESA
Business process mapping	It is a mapping between business capability and business process (or business activity)	[PRM]
IT service mapping	It is a mapping between a capability and an IT service (functional service, operational service, microservice, etc.)	[SRV]
Organization mapping	It is a mapping between a capability and an organizational role	[ORG]

Table 5-6. *Property Attributes of ESA Capability Element*

Category	A-ESA Property	Description
Assessment	Current issues and challenges	Summarized from IT strategic plan
	Improvement opportunities	Summarized from IT strategic plan
	Maturity	– Level 0: Nonexistent – Level 1: Ad hoc – Level 2: Basic (more are in progress) – Level 3: Defined (industry norm) – Level 4: Optimized (differentiating for the organization) – Level 5: Progressive (leading practice in the industry, very few though)

(continued)

Table 5-6. (*continued*)

Category	A-ESA Property	Description
Diagnostics	Urgency	High I Medium I Low
	Importance	High I Medium I Low
	Priority	High I Medium I Low
	Weight	For example, between 10 – 100%
	Decision	New I Retire I Enhance I Preserve
	Purpose	– Foundational Capability – Business Capability – Information Technology Capability – Strategic Capability – Enterprise Portfolio Capability – Digital Capability
Scoring	Effectiveness	For example, rating between 1 – 5
	Controllability	For example, rating between 1 – 5
	Economy	For example, rating between 1 – 5
	Synergy/collaboration	For example, rating between 1 – 5

(*continued*)

Table 5-6. (*continued*)

Category	A-ESA Property	Description
Content	Subgroup	– IT Services or Components – Systems – Business Processes – Business Activities – Generic Services – Course of Action – Resources – Information – Role – Relationship/Interfaces (input and output) – Collaboration Issues – Objectives/Responsibilities

Note:
– Although a capability element can have contain lots of subset attributes, be sure to focus on only a few that matter more to the solution.

In general, the capability model serves two purposes: 1) capability structure and relationship mapping to business and IT architecture domains, and 2) heat mapping.

The heat mapping can be represented with different capability colors/shades based on the ratings of various criteria, such as cost, revenue, IT coverage, performance, strategic importance, urgency, and so on. For example, a typical heatmap uses the *scoring* criteria in Table 5-6.

The components of a capability model are up to each solution needs with a varying degree of detail. Many mappings in the ESA capability model can be in a *many-to-many* relationship, but a less dependent, modular approach is recommended.

Remarkably, capability modeling involves platform thinking. It's not a formality to go through, but rather a critical modeling view, especially when getting into the solution level, to see how different services come together and how the *shared capabilities* (even the utility capabilities) among them. Without this capability thinking, the architecture tends to become more complex as it grows.

HINT

Where Do You Identify Enterprise Capabilities? For better alignment, enterprise capabilities can be identified from a *top-down*, *bottom-up*, or *meet-in-the-middle* perspective. The capability model usually starts with the value chain, which includes all the steps and processes for an enterprise products and services.

Mapping with Value Stream

A *value stream* is the set of actions from initial request to customer value realization. See Figure 2-9 in Chapter 2 for a mapping process between capabilities and value streams. Table 5-7 shows a basic mapping between value stream concepts and ESA notations.

Table 5-7. *Mapping Value Stream to A-ESA Notation*

Category	Value Stream Concept	A-ESA	Notes
Value Process	Value stream	[VL], [PS]	
	Stage	[UC]	Depends on the context of its use
		[TK]	
		[GS]	
Value Criteria	Productivity	[RQ-NFR]	Includes output and input
	Efficiency	[RQ-NFR]	Includes actual output or effective capability
	Flow time	Property attribute	
	Queue/Wait time	Property attribute	
	Resource utilization	[RQ-NFR]	
	Cost	Property attribute	Checks against value [VL]
	Total labor efforts	Property attribute	
	Activity ratio	Property attribute	
	Cycle time	Property attribute	
	Lead time	Property attribute	
	Completeness	Property attribute	
	Accuracy	Property attribute	
	Value-creating time	Property attribute	
	Performance metrics	[RQ]	
	Data integrity/compensation	[DS], [PTN]	
	Work in process	Property attribute	
	Throughput volume/time period	[RQ-NFR]	

Value streams can also be modeled using other frameworks.[8] For example:

- *SAFe* has two similar classes of value streams: operational and development. They can be mapped using the same approach.

- *SIPOC* (Suppliers, Inputs, Processes, Outputs, and Customers) from Lean Six.

- *Sigma* is a lean modeling concept of processes that enables businesses to assess the efficiency of each business capability instance in terms of waste through lead-time and throughput time.

- *Journey maps* can align a map with the value stream stages and the underlying capability instances. As shown in Table 5-8, the journey map extends the value stream by evaluating the customer experience, and its touch point is equated to the value stream stage. This add-on incorporates customer feedback into the capability model for improvement.

Table 5-8. *Mapping Between Value Stream and Journey Map*

Persona	Customer			
Scenario	Using Order Process App			
Touch point	UC Access OPS App	UC Browse Shopping Cart	UC Order Placement	UC Order Fulfillment
User feedback	OK	OK	Peak time congestion	OK
Capability instance improvements	None	None	Need to improve OPS scalability	None

---

[8] Reference: *Business Capability Planning,* The Open Group

Style 4: Business Architecture

In A-ESA, business architecture is more focused on the relationship to solution contexts and case scenarios.

Relationship Between Business Model and Business Architecture

The business model is designed to assist stakeholders in evaluating the overall feasibility and in making strategic decisions. However, it typically does not include sufficient information to implement the IT strategy. What is crucial in A-ESA is the alignment of the business architecture with the IT strategy. The following is a brief mapping between the business model (a skeleton outline) and the business architecture (a simple mapping with A-ESA notations).

The process begins with an analysis of *Porter's Five Forces* (see Table 5-9), which provides insight into the strategic position. This is followed by a framework analysis of the business model (see Table 5-10), and finally, a mapping of the business architecture (see Table 5-11).

Table 5-9. *Porter's Five Forces*

Threat of New Entrants [RO]	Bargaining Power of Suppliers [RO]	Bargaining Power of Buyers [RO]	Threat of Substitute Products [PD]
	↓		
Rivalry Among Existing Competitors			
Note: – For an example case, see https://www.business-to-you.com/porters-five-forces/.			

Table 5-10. *Nine Area Business Model Canvas*

Key Partners [RO] (e.g. payment providers)	Customer Segments [DM] (e.g. niche market)	Customer Relations [GS] (e.g. self-service)
Value Proposition [VL] (e.g. high performance)	Channels [UI] (e.g. mobile shop store)	Key Activities [TK] (e.g. software development)
Key Resources [CP] (e.g. IT human resources)	Revenue Streams [VL] (e.g. AD revenue)	Cost Structure [RQ-CST] (e.g. platform cost)

Note:
– For an example case, see https://www.retaildogma.com/retail-business/.

Table 5-11. *Business Architecture Mapping*

Business Architecture Blueprints				
Operating Models [ML]	Value Chain Models [PRM]/[VL]	Supplier Chain Model [CAP]	Value Streams [VL]	Dashboards/ Balanced Scorecard [KC]
↓				
Business Architecture				
Organization Map [ORG]	Capability Map [CAP]/[VL]	Case Scenario and Process Decomposition [UCM]/[PRM]	Business Service Mapping [PRM]	Information Mapping [IRV]

Table 5-12 shows the mapping between business architecture (BA) and ESA. The overarching business model is ultimately mapped to the business architecture model, as illustrated in Profile 2.

Table 5-12. *Mapping Between BA and ESA*

BA	Mapping BA to ESA	Description	A-ESA
Information Map	Data Architecture	Information map (generally containing subject areas and data items) is detailed into data architecture (containing data entities), and realized from information view (containing data services and potentially data domains).	– Information View [IRV]
Value Map	Process & Case Management Automation	Value map is orchestrated via process notation, say, BPMN (business process model and notation), and realized from process view or page flow view, or value stream can be mapped to capability via value stream stage and realized from application element.	– Value Stream [VL]

(*continued*)

Table 5-12. (*continued*)

BA	Mapping BA to ESA	Description	A-ESA
Capability Map	IT Capability	Business capability map is transformed to IT capability, outlined by generic services and realized by functional services, or the capability can be mapped to an application element containing a group of coherent services.	– Generic Service [GS] – Application [AP] – System [SY]
Interaction Model	User Interface and Use Case Interaction	Interaction model deals with user experience and interaction between business and application.	– Use Case View [UCM] – Page Flow View [PFV]
Process Model	Process Decomposition and Optimization	Business process is mapped to application process or use case model, which ultimately triggers functional services or microservices.	– Process View [PRM] – Use Case View [UCM] – Process Service [PS]

Note:

– In a complex service-oriented architecture, there is a need for operational view mapping to integrate technical and middleware services.

Organization Communication Structure

As shown in Table 5-13, the organization that is part of the business architecture is typically structured in close or partial alignment with its high-level business capabilities. It may also include geographic locations that are limited to the business capability catalog.

Table 5-13. Partial Retail Organization

Department	Product	Human Resources	Logistics	Sales
Function	Product Development [RO]	Training Group [RO]	Transportation [RO]	Online Store [RO]
	Vendor MGT [RO]	Office Support [RO]	Demand Planning [RO]	Retail Store [RO] – San Jose, CA [LO] – Houston, TX [LO]
	Sourcing [RO]	Compensation & Benefits [RO]	Delivery MGT [RO]	

The organizational structure serves as an input to A-ESA. In light of Conway's Law, which posits that the communication structures of an organization influence its solution architectural styles and development environment, it is imperative that solution architects consider the organizational factors at play. A brief summary of the organizational information related to the solution architecture can be selectively included in the modeling mapping, as seen in a simple illustration in Table 5-14 .

Table 5-14. *A-ESA Mapping to Organization Structure*

Example Metric	Example Property Attribute	A-ESA
Organization Maturity	1. Initial 2. Managed 3. Defined 4. Quantitatively managed 5. Optimizing	Organization View [ORG]
Development Team Style	– Structured – Agile	Organization View [ORG]
DevOps Role	– Scrum master – Product owner – Technical support – Architect – Analyst – Software engineer – Tester	Role [RO]
Stakeholder's Power of Influence	– Jack Luwinski (decision-maker) – Agies DevGoots (innovation-minded) – Odea Timmel (OPEX[9] advocate)	Role [RO]

The organization's IT strategy in part shapes its capabilities, structure, collaborative processes, people, and rewards, which are related to how people are shaped by goals. Table 5-15 shows an ESA mapping to these organizational focus considerations.

---

[9] OPEX (operational expenditure) is the money a company spends on an ongoing, day-to-day basis to run a business or system (whatis.com).

Table 5-15. *Organization Focus Mapping with ESA*

Organization Focus	Inclusion Description	A-ESA Notation
Capability	Business capabilities, technologies, or processes	[CAP]
Structure	Functional, product, geographic, customer, or hybrid	[ORG]
Process	Collaboration among cross-business teams	[PRM]
People	Integrative role	[RO]

Note:

– Customer-centric digital organization is attributed to Agile, transparent, iterative, innovative, etc.

– The organization pretty much determines the IT architectural styles. For example, the flexible team structure and the flexible Internet product lines are good for Agile development process.

Organization design (OD) is thought of as part of the *Target Operating Model* (TOM),[10] of which the solution operating model is a part. If an organization is cross-functional, it's natural and likely to have a DevOps operating model where DevOps teams are responsible for development and deployment, while platform teams are responsible for development platform support, operational environment design, network design, operational management, and backup and disaster recovery, and so on.

HINT

Dynamic Organigram: The organization's structural map can also be represented in the form of a web-style *organigraph* without adhering to a more stable linear hierarchy for dynamic enterprise ecosystem relationships between departments, stakeholders, partners, products, locations, and supply chains that may not otherwise be apparent.

---

[10] TOM is an enterprise strategic framework in various forms and generally outlines how people, organization, processes, technology, and governance should interconnect to provide value.

Business Objectives

A-ESA is primarily concerned with non-functional requirements. These requirements are partly derived, explicitly or implicitly, from strategic directives and business requirements. As seen in the previous chapters, requirements mapping is a key modeling activity. Table 5-16 shows a mapping illustration of a typical business terminology. This is a similarity analysis that requires working with the solution architect for consensus.

Table 5-16. *Mapping Fuzzy Business Term with NFRs*

Business Objective	NFRs	A-ESA Notation
Flexibility	Composibility, modularity, reusability	Service-based or component-based architecture [AOV]
	Agility	DevOps, Agile modeling [DEV]
	Elasticity, resilience, scalability	Circuit breaker, traffic control, self-healing, virtualization [DEP]
Note: – There are more NFRs related to flexibility such as customizability, configurability, deployability, dynamicity, changeability, extensibility, and adaptability.		

The real intent and feasibility need to be further nailed down through the modeling and architecture decision process. A good practice is to limit the NFR terms to a few salient ones for each solution, for example, using the term *elasticity* to imply *scalability* and *resilience* when the concurrent burst is the main concern.

Business Scenario

A case scenario covers the *problem space* and the *solution space*. An *intent-driven* business scenario is more in the problem space. The A-ESA case scenario maps the problem space but also considers the solution space. All case scenario requirements are embodied by A-ESA elements or views. Table 5-17 shows a quick comparison.

Table 5-17. *A-ESA Mapping to Business Scenario*

Category	Subject	Business Scenario	A-ESA Scenario	A-ESA
Assessment	Roadmap	More EA coverage	More SA coverage	[CAP], [AOV]
	Priorities	Heat maps	Hotspots, significant scenarios (walkthrough)	[DEP]
	Evaluation	Gap analysis	Architecture decisions	[KC], [VF]
	Alternative weighing criteria	Goal, cost, revenue, resource, scope and timeline	Principles, Service-level requirements, tradeoff analysis within the restrictions	[PR], [RQ-NFR], [KC]

(continued)

Table 5-17. (*continued*)

Category	Subject	Business Scenario	A-ESA Scenario	A-ESA
Context	Rationale	More on the why	More on the how	[MTS]
	Stakeholder	Who will use	Who will use it and how	[RO]
	Background	Business driver	Solution needs	[AOV], [ORG], [CAP]
	Problem statement	Issue, pain point	Requirement, risk	[RQ], [RK], [GV]
	Business environment	Constituencies	Input mapping	[AOV] (System Context), [PRM], [UCM]
	Technical environment	Conceptual	Logical and physical	[AOV], [DEP]
Content	Principles	Guiding principles	Architectural principles, solutions principles	[PR]
	Business process	Value stream	Business application process	[VL], [PRM]
	Requirement	General Requirement	Functional Requirement, non-functional requirement, rules, governance	[RQ], [RQ-RL], [GV]
	Relationship to processes	Intent-oriented	Results-oriented	[SRV]
	Architecture model	Business model	Solution model	[ML], [VF]
	Outcomes	Intended outcome	Deliverable	[DL]

Style 5: Information Architecture

Information architecture contains information assets or processed data and their relationships. Table 5-18 shows how to express common information or data relevance using A-ESA notations.

Table 5-18. *A-ESA Mapping to Information Architecture*

Category	Information Assets	A-ESA Notation
EA	Information mapping to capability	Data Service [DS]
	Data and information compliance	Governance [GV]
	Data guidelines	Principle [PR]
BA	Analytical data	Data Service [DS]
	Process data	Data Service [DS]
	Business object	Object [OB]
	Data table	Data Object [TB]
	Data entity	Data Object [EN]
	Value stream	Value Stream [VL]
	Value stream map	Process [PS]
	Business service	Generic Service [GS]
	Business function	Task [TK], Use Case [UC]
	Business activity	Activity [TK]

(continued)

Table 5-18. (*continued*)

Category	Information Assets	A-ESA Notation
Structural	Unstructured data, such as documents,	Data Service [DS], Document File [DF]
	Structured data such as Markup language (XML, etc.), message data	Data Service [DS], Message Queue [MQ]
	Information including catalogues, ontologies, taxonomies, metadata, templates, etc.	Document File [DF], Repository [AR-RP]
	Database schema	Data Schema [SM]
Operational	Database	Data Store [DB]
	Data warehouse, data lake	Data Store [DB]
	Data integration and interoperability	Service Interface [SI]
	Data transaction	Domain [DM]
	Data quality, data security	Requirement [RQ-NFR]
	Operational data	Data Service [DS], Operational View [DEP]
	Data object	Data Object [OB]
	Session data	Session [SS]
	Caching data	Caching [CA]
	Big data, MDM (Master Data Management)	Generic Service [GS], Pattern [PTN]
	Data event	Event Service [ES]
	Granular data service	Data Object [OB]

Note:

– Business intelligence and analytics are part of information architecture.

– The A-ESA elements in the table cover the DMBoK (Data Management Body of Knowledge) DAMA Wheel.

Information architecture is generally a high-level view of enterprise, business, structural, and operational architecture. As shown in Table 5-18, it also includes a high-level view of data architecture. While solution-level information architecture focuses on the design and usability of information from the user's perspective, data architecture focuses on the technical management and governance of data from the system's perspective. So, no matter how you divide it between information and data architecture, there should be a smooth flow between information and data architecture.

HINT

Data Governance: Data governance is a critical component of information architecture and data architecture, starting from information architecture. The goal of data governance is to enhance the value of data. Data governance is composed of management systems and technical systems, including organizations, systems, processes, technologies, and supporting tools. It is closely related to compliance frameworks, data quality, and security in ESA.

Style 6: Application Architecture

The application architecture overlaps with the functional architecture. Loosely speaking, application architecture encompasses the functional architecture and part of the DevOps environment.

Functional Architecture

Table 5-19 shows typical A-ESA elements used in functional architecture (a core part of A-ESA). Functional SLC services play a key role in a well-formed solution architecture.

Table 5-19. *Typical Functional Architecture Elements*

Category	Functional Services	Description	A-ESA
Logical or functional SLC service	User interface	An interactive service, generally with a visual presentation	[UI]
	App logic service	Explicitly defined non-GUI application behavior and control logic	[AS]
	Data service	A self-contained piece of information including a standalone data object, or a federated, integrated piece of data service	[DS]
	Technical service	A technical behavior, independent of the application-specific logic context	[TS]
	Generic function service	An independent service that contains UI, AS, DS, or TS	[AP], [GS], [MS]
Implementation service	Service interface	Service interface, contract, API	[SI]
	Service component	Application component, technical component	[SC]

279

In ESA, application architecture means more functional architecture and involves multiple tasks/steps and multiple views. Critical to application architecture are the *domain* and/or service-oriented architecture processes.

DevOps Environment

DevOps has a lot to do with the application architecture and the development process. It involves functional and operational aspects, along with cultural considerations. Table 5-20 shows the A-ESA elements in a DevOps environment.

Table 5-20. *DevOps Environment Mapping with A-ESA*

DevOps Environment	Description	A-ESA
Organization & Culture	Organization structure and innovation culture	[ORG]
Process	Continuous integration and continuous deployment process	[PS]
Tooling	Automation tool chain	[MW]
Deployment	Mapping functional services to deployment packages	[DPM]
Note: – DevOps is closely related to architectural governance.		

Style 7: Data Architecture

Data architecture is often confused with information architecture. In ESA, information architecture is more closely related to business architecture and enterprise planning, while data architecture is more closely related to application and technical architecture. Data architecture is considered as an integral part of the solution operating model.

Data architecture expresses data structuring, data processing and collaboration, data transactions, data distribution, data access, and data storage. It is the foundation for data integration operations and artificial intelligence (AI) applications.

Data Transaction Considerations

Data transaction mechanisms are the most important consideration in most data solution architectures. Depending on the requirements and the solution governance, different transaction mechanisms are considered. Table 5-21 provides a list of common transaction patterns.

Table 5-21. *Data Transaction Considerations*

Consistency	Transaction	Description	A-ESA Notation
Strong consistency	2PC	Two-phase commit/XA is a highly consistent transaction, but there are still risks such as data inconsistency and blocking. It is only applicable at the database level.	– Transaction [TR] – Event Service [ES] – Message Queue [MQ] – Table [TB] – Database [DB] – Pattern View [PTN]
	3PC	Three-phase commit includes CanCommit, PreCommit, and DoCommit. It has similar mechanisms to 2PC.	
	TCC	Try-Confirm-Cancel is a compensatory transaction idea that has a broader scope of application and is implemented at the business level, so it is more intrusive to the business.	
Weak consistency	Sagas Pattern	It splits the transactions and is a compensation mechanism.	
	Local Message Table	It relies heavily on the message table of the database to manage transactions. It is generally rarely used for high concurrency scenarios.	
	Message Transaction	It implements transactions directly on top of MQ.	
	Best Effort Notification	It is an eventually consistent transaction.	

Eventual consistency is unlikely to be specified as an explicit requirement of a distributed system. Instead, it is often a consequence of implementing a system that must exhibit scalability and high availability, which precludes most common strategies for achieving strong consistency.

For data consistency, there may need to be a wide range of alternatives to make an appropriate decision. In general, you have the choice of *strong consistency [ACID]* or *weak eventual consistency [BASE]*. Table 5-22 provides a list of common transaction considerations.

Table 5-22. *Transaction Considerations*

Consideration	Decision Choice	A-ESA
Strong consistency	Linearizability consistency	[PTN]
Strong consistency	Sequential consistency	
Weak consistency	Causal consistency	
Weak consistency	Eventual consistency	
Business affability	1. XA (choice) 2. SAGA 3. TCC	[KC]
Assured consistency	1. SAGA 2. TCC (choice)	
Degrading Performance	1. Message transaction 2. TCC (choice) 3. SAGA	

Note:
— For an overall considerations of business affability, consistency, and performance, XA is probably the better choice. Consistency for SAGA depends on the business rules. It's not a guaranteed consistency.

In a distributed architecture, data consistency is a must decision. In many situations, an ACID consistency is not necessary. The degree to which consensus must be reached is up to the specific case scenario. For example, weak consistency is more appropriate for a gaming platform or a social media platform like Spotify where teams are autonomous.

In a public distributed database environment, there is a choice of vendor-defined consistency levels (based on quorum[11] reads/writes) and runtime modes to meet data durability or *recovery point actual* (RPA) requirements.

HINT

Viewpoint of Data Consistency: Consider data consistency from a user viewpoint: monotonic read consistency, monotonic writes consistency, read your writes consistency, and so on. In addition, view consistency from a data viewpoint: causal consistency, sequential consistency, and so on.

AI Considerations in Data Architecture

Typical AI architecture is closely related to *big data architecture* (see Solution Profile 8), which requires data federation, data quality, data management, data security, and compliance. Big data leverages AI for better data analysis. In turn, AI requires a gigantic amount of data to learn and improve from big data analysis. Table 5-23 shows some typical AI data considerations.

---

[11] A *quorum* is the minimum number of votes that a distributed transaction has to obtain.

Table 5-23. *Rough A-ESA Mapping for AI in Data Architecture*

Category	AI Concept	A-ESA Notation
Activity	Model selection such as: – Interference mode – Batch processing mode – Stream processing mode – Online/offline learning mode – Data labeling mode	Key Choice [KC]
	Ethical considerations	Key Choice [KC]
	Data scoring	Task [TK]
	Data model training	Task [TK]
	Hyper parameter tuning	Service Interface [SI], Task [TK]
Case Scenario	AI data usage scenario	Use Case [UC], Requirement [RQ]
Pattern	Machine learning reference architectures	Pattern View [PTN]
	Pre-built AI models in AI Builder	Model [ML]
Information Service	Machine learning/knowledge engineering	Data Service [DS], Information View [IRV], Artifact [AR]
	Deep learning/cognitive computing structuring and design	Data Service [DS]
API	Data access via prebuilt AI, OpenAI API, LLM API, GitHub Copilot API	Service Interface [SI], Cloud [CL]
Development	AI data service implementation, version control, and testing	DevOps [DEV]
Operational performance	Data allocation, performance, efficiency, scalability, etc.	Node [ND], Database [DB], Data Service [DS]

Note:
– See Table 5-51 for a representative AI reference architecture.

Note that AI data considerations focus on data preprocessing, data model training, development, API design, testing, version control, optimal machine learning tuning and performance, convergence of databases and AI, data resource allocation, along with ethical constraints and scalability and flexibility requirements. Unlike AI, data architecture is more concerned with data storage, data integration, data governance, data access and indexing, data lifecycle management, data backup, and disaster and recovery.

The integration of AI into data architecture can be a highly effective strategy, as evidenced by the success of Airbnb's recommendation system. By leveraging machine learning algorithms and big data analytics, Airbnb has been able to implement dynamic pricing and personalized recommendations, enhancing listing visibility and boosting booking rates.

HINT

Is AI an Arch Style? As you know, the hot and trendy AI relies heavily on three amigos: deep learning algorithms and LLM (large language model), data and computing power. Traditionally, AI fell under the base data architecture or data analytics reference architecture. Now, the base technical architecture is playing an increasing role in AI. To many people, AI is an application innovation that requires design thinking. AI is an interplay of several disciplines beyond IT architecture. AI architecture is more about AI application, data manipulation and analysis, and its basic architecture is intertwined with other architectural styles. Considering its nature of comprising relatively independent architectural types, AI is generally regarded here as a high-level reference architecture.

Style 8: Technical Architecture

In general, technical architecture covers *functional* and *operational* aspects. A-ESA technical architecture tends more toward operational architecture. It is also more toward reference architecture, patterns, tradeoffs, operational SLA, as well as its correlation with application architecture.

Operational Architecture

Table 5-24 shows typical A-ESA elements used in an operational architecture (a core part of A-ESA).

Table 5-24. *Typical Operational Architecture Elements*

Category	Op Environment	Description	A-ESA Notation
Location	Location	Including geographic location or physical location of a facility, data center, etc.	Location [LO]
	Domain/Scope	Including tier, region, zone, cell, environment, platform, cloud, cluster, transaction, microservice, etc.	Domain [DM]
	Layer	Logical group of representations	Grouping [GP]
Node	Node	Physical computing nodes	Node [ND]
	Virtual Node	Virtual container	Virtual Service [VS]
	System	Systems or devices	System [SY]
	Network Device	Including hardware-level proxy, firewall, switches, bridges, etc.	Network [NW]

(continued)

Table 5-24. (*continued*)

Category	Op Environment	Description	A-ESA Notation
Containing Unit (on the node)	Middleware	Technical product or off-the-shelf software ready for use through configuration	Middleware [MW]
	Deployment Unit	A group of functional services that share the same SLCs defined from the solution	Deployment Package [DP]
Link (of the node)	Connection Relationship	Includes connections of various types: concurrency, speed, bandwidth, intermittency, etc.	Directional Relationship [FW], Association Relationship [AN]

Technical Platform Architecture

Bespoke technical architecture often leads to over-complication. Therefore, changing the "in house architecture" mindset is an issue for the technical platform. Using reference architectures and patterns is a good way to mitigate technical challenges. Two representative technical platform architectures in an operational environment are presented in this section.

Distributed Transactional Architecture

Microservice architecture shifts complexity to the lower layers. It requires a modern distributed transactional architecture to ensure eventual data consistency while achieving high availability and scalability.

Table 5-25 showcases a distributed transactional architecture platform that handles simultaneous application updates from multiple users to address key operational concerns: performance, scalability, elasticity, and reliability. It is a technical platform architecture with data architecture considerations.

Table 5-25. *A-ESA Mapping with Distributed Transactional Architecture*

Layer	Description	A-ESA Notation
Public Cloud Environment	An environment for redundant network communications across multiple sites organized into regions and zones, typically provided by a large public cloud with the right infrastructure and messaging services.	Cloud [CL] Domain [DM] Zone [ZN] Region [RG] Network [NW] Node [ND]
Microservices	Independently deployable service units.	Microservice [MS] Deployment Unit [DP] Key Choice [KC]
	FaaS in a serverless environment.	
Docker Containers	A containerized virtual environment that includes a minimal OS image and associated dependencies and libraries.	Virtual Node [VN]
Container Orchestration	A way to manage the containers' SLCs. It provides a pod, which often includes sidecar containers with common utility services for improved modularity.	Process [PS] Pod [DM-PO)
Distributed Database Services	A persistence layer that ensures low-latency performance and overall consistency in a cloud environment.	Database [DB] Key Choice [KC]

There are many options for a distributed transactional architecture. The choice of deployment approach (serverless or dedicated deployment) or distributed database depends on your architecture decision in terms of cost, resource utilization, self-control, manageability, and so on.

Cell-based Architecture

Table 5-26 illustrates a *cell-based architecture* (CBA) mapping with A-ESA model building blocks at the domain-level using a system-of-systems approach. It reflects the transitions between domain-level considerations and the technical platform environment.

Table 5-26. *A-ESA Mapping in Cell-based Architecture*

CBA Notation	Description	A-ESA Notation
General		
Principle	Cell-based architecture has four principles: scalability, modularity, composability, and governance.	Principle [PR]
User	A human or an actor interacting with the system.	User [RO]
Component	The atomic unit of this reference architecture.	App Logic Service [AS]
Cell	An architecture block.	Cell [DM-CE]
Component Types		
Legacy and data services	Databases, existing systems, registries and repositories, user stores, and business processes.	Database [DB] System [SY] Process [PS]
Microservices and serverless components	Custom business logic, aggregation and service composition, and transformation.	Microservice [MS] App Logic Service [AS]
Gateways and brokers	Events, streams, policy enforcement points, message brokers, identity brokers and identity gateways, sidecars, and bridges.	Edge Service [EG] Event Service [ES] Message Queue [MQ] Middleware [MW]
External endpoint	Access using APIs, streams, cloud systems, and SaaS.	Service Interface [SI] System [SY] Cloud [CL]
Frontend clients	Mobile apps, web apps, portlets, reactive apps, and API consumers.	Frontend Client [FE]
Governance and utilities	Registry, observability, automation tools, and lifecycle management.	Governance [GV] Middleware [MW]

(*continued*)

Table 5-26. (*continued*)

CBA Notation	Description	A-ESA Notation
Inter and Intra Cell Communication		
Data plane	It routes traffic between hops and accepts data packets.	Data Service [DS]
Control plane	Signaling of the network, makes decisions about the traffic flow, runtime governance.	App Logic Service [AS]
Management plane	Configuration, observability, monitoring, and design-time governance.	Tech Service [TS]
Domain-level Cell Type		
Logic	Microservices, functions, micro-gateways, and lightweight storages.	Generic Service [GS] Microservice [MS] App Logic Service [AS]
Integration	MicroESB or other integration microservices, lightweight storage, and/ or caches.	Tech Service [TS] Edge Service [EG]
Legacy	Existing systems, legacy services, and COTS systems.	System [SY]
External	SaaS and partner systems.	Service Interface [SI] Cloud [CL]
Data	RDBMS, NoSQL, files, message brokers.	Data Service [DS] Message Queue [MQ]
Security	IDP (identity provider) and user stores.	Database [DB]

(*continued*)

Table 5-26. *(continued)*

CBA Notation	Description	A-ESA Notation
Channel	Web, mobile, and IoT (end-user) applications.	User Interface [UI] Mobile Device [MD]

Note:
- Reference source: *Cell-based Architecture: A Decentralized Reference Architecture for Cloud-native Applications* https://github.com/wso2/reference-architecture/blob/master/reference-architecture-cell-based.md.
- The "onion structure" in a cell-based architecture is a design pattern that organizes software components into concentric layers and aims to separate business logic from infrastructure concerns.

Cells are built to be flexible and scalable so they can handle cloud-based apps. For example, in a service-based architectural framework, you can introduce an alternative reward system either outside the service cluster or as a secondary server within the core service collection. With the cell-based approach, it's pretty straightforward: just create a new cell, specify the brokered communication, activate the various gateways and brokers, and activate the cell. This is because everything is abstracted into unique entities.

Physical Architecture

Typically considered a part of the technical architecture, the physical architecture must not only provide the infrastructure, but also ensure high performance, reliability, security, scalability, and cost-effectiveness of the system. At the same time, it must be closely aligned with operational and technical strategies to support the overall goals and business needs of the enterprise. Therefore, when considering the operational architecture, always consider servers, storage, and network devices, as well as related NFRs such as disaster recovery.

Note that the physical substrate of the technical architecture is not a major ESA concern, as it is *relatively stable*. ESA serves as a blueprint for the IT infrastructure and leaves the design to the infrastructure domain expert.

Style 9: Service-oriented Architecture (SOA)

SOA is passé. It is an *architectural style* by intent, but mistaken by many for Web services, technical specifications, or assets. A-ESA is an *IT service-based architecture* that incorporates SOA services and MSA[12] services, as well as enterprise solution environment considerations.

SOA is based on the design of solutions using services. Here the A-ESA mapping is more like the IT service mapping. It provides consistent abstractions from high-level strategies and deliverables to architectural solutions. SOA in an A-ESA model uses lightweight guardrails. Table 5-27 shows a mapping between A-ESA and SOA reference architectures.

---

[12] Microservice Architecture

Table 5-27. *SOA Reference Architecture Mapping to A-ESA*

Area	SOA Layer	A-ESA Element	A-ESA View
Strategic Input	Strategic Capability	– Capability [CP] – Principle [PR] – Role [RO] – Generic Service [GS]	– Capability View [CAP] – Organization View [ORG]
Solution	Consumer Interfaces	– Role [RO] – GUI Service [UI]	– Page Flow View [PFV]
	Business Processes	– Process Service [PS] – Use Case [UC] – Task [TK]	– Process View [PRM]
	Services	– Domain [DM] – User Interface [UI] – App Logic Service [AS] – Tech Service [TS] – Data Service [DS] – Service Interface [SI]	– Functional View [SRV]
	Service Components	– Service Component [SC]	– Service Component Realization [SCR]
	Operational Systems	– Deployment Package [DP] Middleware [MW] – System [SY] – Node [ND] – Network [NW] – Location [LO] – Virtual Node [VN]	– Operational View [DEP]

(continued)

Table 5-27. (*continued*)

Area	SOA Layer	A-ESA Element	A-ESA View
Cross-cutting Concerns	Integration	– Middleware [MW] – Key Choice [KC] – Domain [DM] – Service Interface [SI]	– Integration View [AOV] – Functional View [SRV] – Operational View [DEP]
	Quality of Service	– Requirement [RQ-NFR], Risk [RK]	– Deployment Package Mapping [DPM] – Operational Walkthrough View [DEP]
	Information	– Data Service [DS] – Data Store [DB] – Data Object [OB]	– Information View [IRV]
	Governance	– Governance [GV] – Key Choice [KC]	– Metrics View [MTS] – Validation View [VLD]

Table 5-28 shows a rough mapping of element notations between SOA and A-ESA.

Table 5-28. *Notation Mapping between A-ESA and SOA*

SOA Element	A-ESA Representation
Business Process	Composite Service/Process [PS]
Class Symbol	Entity [EN]
Client Workstation	System/Device [SY], PC Device [PC]
Component	Service Component [SC]
Component-based Service Contract	Service Component Interface [SI]
Conflict	Risk/Issue [RK]
Database	Data Store [DB]
Design Principle	Architecture Principle [PR]
Generic Application	Application [AP], Generic Service [GS]
Grid Service	Domain [DM], Network [NW], Middleware [MW]
Human	Role [RO]
Human Readable Document	Document File [DF]
Message	Message Service [MG]
Message Queue	Message Queue [MQ]
Physical Container	Node [ND]
Runtime Processing	Deployment Package [DP]
Service	Generic Service [GS], Microservice [MS]
Service Agent	Edge Service [EG]
Service Composition	Process Service [PS], App Logic Service [AS]

(continued)

Table 5-28. (*continued*)

SOA Element	A-ESA Representation
Service Contract	Service Interface [SI]
Service Inventory	Capability [CP], Repository [RP]
Service Inventory Container	Deployment Package [DP], Virtual Service [VS]
Service Layer	Domain [DM], Group [GP]
Service with State Data	Object [OB], Cache [CA]
Strategic Goal	Requirement [RQ-GL]
Transition Arrow	Flow Relationship [FW]
User Interface	GUI Service [UI]
Web Service	Service [GS], Tech Service [TS]
Web Service Boundary	Service Interface [SI]
Web Service Intermediary	Edge Interface [EG], Tech Service [TS]
WS Policy Definition	Governance [GV]
WSDL Definition	Governance [GV]
XML Schema Definition	Governance [GV]
Zone	Zone [DM-ZN]

Style 10: Microservice Architecture (MSA)

Microservice architecture requires a lot of architectural expertise. MSA is a more sophisticated architecture, so this section describes a bit more about the MSA style.

Microservice Architecture Principle

First and foremost, an MSA application must follow a set of principles, commonly known as the five principles: 1) single concern, 2) discrete, 3) transportable, 4) carrying its own data, and 5) inherently ephemeral. A-ESA follows five microservices principles in a solution context (see Table 5-29).

Table 5-29. *MSA Principles*

Architecture Area	Principle	ESA Notation
Business	Model around business domain	– Capability View [CAP] – Process View [PRM] – Architecture Outline View [AOV]
Application and Data	Design independently deployable services	– Service Relationship View [SRV]
Application and Development	Cherish DevOps and automation	– DevOps View [DEV]
Technical Runtime	Operate in a resilient and scalable environment	– Cloud [CL] – Middleware [MW] – Virtual Node [VN]
	Manage in an observable environment	– Middleware [MW]
Note: – The logical architecture and logical boundaries of a system do not necessarily have a one-to-one mapping.		

Microservice Architectural Process

Microservice architecture is a scalable architecture suitable for Agile and digital transformations. It is often a project or product level architecture, not a full enterprise architecture per se. Table 5-30 shows a loosely-defined microservice architecture process that maps A-ESA views and key elements.

Table 5-30. *ESA-level Microservice Architectural Process*

Area	Layer	A-ESA View	A-ESA Element
Design Thinking	Value and requirement mapping	– Process View [PRM]	– Value [VL] – Generic Service [GS] – Use Case [UC] – Process Service [PS] – Task [TK]
	Domain-driven design (DDD)	– Use Case Scenario [UCM] – Architecture Overview [AOV]	– Intent-oriented Domain [DM] – Bounded Context [BC]
Frontend Engagement	Modern internet (mobile app, single-page app)	Page Flow View [PFV]	– User Interface [UI]
	Page and data caching	Data Flow View [IRV]	– Data Service [DS]
Microservice Functional Architecture	Backend for frontend (BFF)	Functional View [SRV]	– Tech Service [TS]
	Functional services	Functional View [SRV]	– App Logic Service [AS] – Data Service [DS] – Tech Service [TS]
	Microservices	Functional [SRV]	– Result-oriented Domain Context [BC] – Microservice [MS]
	Microservice components	Component Realization [SCR]	– Service Component [SC]

(continued)

Table 5-30. (*continued*)

Area	Layer	A-ESA View	A-ESA Element
DevOps Environment	Organization structure	Organization [ORG]	– Role [RO]
	DevOps chains	DevOps [DEV]	– Middleware [MW]
	Microservice runtime	Deployment Mapping [DPM]	– Deployment Unit [DP]
	Microservice registration	DevOps [DEV]	– Service Interface [SI] – Middleware [MW]
	Logging aggregator (Correlation ID)	DevOps [DEV]	– Key Choice [KC]
Scalable Storage	Caching	Operational [DEP]	– Middleware [MW] – Cache [CA]
	Adaptation services	Operational [DEP]	– Edge Interface [EG]
	Scalable store (key-value, document, or other cloud storage)	Operational [DEP]	– Data Store [DB] – Key Choice [KC]
Cross-cutting	Governance	Metrics [MTS]	– Governance [GV]
	Transactional pattern	Pattern [PTN]	– Transaction [DM-TR] – Data Service [DS]

Microservices Architecture Design

Microservices architecture is a subset of service-oriented architecture that is business-driven, so business architecture, application architecture, and data architecture play a major role. From a business application perspective, MSA follows domain-driven design (DDD), which provides guidance for specifying bounded contexts as services. Table 5-31 shows the core modeling activities for a microservice architecture design. See Chapters 2 and 6 for the corresponding microservice architecture examples and techniques.

Table 5-31. *Microservice Architecture Activities and A-ESA Mapping*

MSA Activity	MSA Term	A-ESA Element	A-ESA View
Intended Domain Analysis			
Domain identification and mapping	Domain	Domain [DM]	Architecture Overview [AOV]
Domain service definition	Domain service	Domain Service [GS]	Functional [SRV]
Resulting Domain Specification			
Domain model specification	Application service	Application Logic Service [AS]	Functional [SRV]
	Domain context service	Bounded Context [BC]	Functional [SRV]
	Aggregate	Entity [EN]	Functional [SRV]
	Value object	Entity [EN]	Functional [SRV]
	Domain event	Event Service [ES]	Functional [SRV]
Microservice boundaries specification	Microservice	Microservice [MS]	Functional [SRV]

(continued)

Table 5-31. (*continued*)

MSA Activity	MSA Term	A-ESA Element	A-ESA View
Microservice data store realization	Data store	Database [DB]	Component Realization [SCR]
		Schema [SM]	Component Realization [SCR]
Microservice Distribution			
Operational pattern and platform	MS reference architecture		Pattern [PTN]
	Runtime microservice	Deployment Unit [DP], Virtual Node [VN]	DevOps [DEV]
	Platform	Node [ND]	Operational [DEP]
		Cloud [CL]	Operational [DEP]
Inter-service communication	Distributed transaction pattern	Transaction Domain [TR]	Pattern [PTN]
	Messaging pattern	Message Queue [MQ]	Pattern [PTN]
	Operational pattern	Service Interface [SI]	Pattern [PTN]
	Service mesh	Middleware [MW]	Operational [DEP]

(*continued*)

Table 5-31. (*continued*)

MSA Activity	MSA Term	A-ESA Element	A-ESA View
Microservice Interface Exchange			
API design	API	Service Interface [SI]	Functional [SRV]
Microservice endpoint and API gateway	API gateway	Middleware [MW]	Operational [DEP]
		Tech Service [TS]	Functional [SRV]

Note:

– Domain refers to a set of data related to a specific purpose. A domain service is an object that implements some logic without holding any state.

– A bounded context is simply the boundary within a domain where a particular domain model applies.

– An aggregate contains root entities that include entities and value objects. An aggregate defines a consistency boundary around one or more entities to model transactional invariants. This must be done in conjunction with the data design team.

– For practical purposes, an entity is also a domain/object that would be used to store in the database.

– A value object is immutable and has no identity. Typical examples of value objects are dates, times, and colors.

– A domain event is something that happens in the domain to be noticed by other parts of the same domain.

– For reference reading, see *Azure Architecture Center - Domain Analysis for Microservices.*

Microservice DevOps

With the widespread adoption of container infrastructure and DevOps practices, resilience-based service architecture is a choice for many solutions. MSA achieves architectural resilience through various techniques (such as circuit breakers, bulkheads, retries, and rate limiters) and various tools, such as Hystrix, Reslience4J, and Sentinel.

Even with the flexible infrastructure and tool support, a solution architect must still deal with the added architectural complexity due to the distributed nature of microservices. Table 5-32 shows a DevOps solution metrics of microservice architecture and associated notations that handle different services by different *priority levels.*

Table 5-32. *Typical MSA Operational Priority Metrics*

Solution Metric	A-ESA	MS-C1	MS-C2	MS-C3	MS-C4
Single point deployment	[GV]			X	X
Deployment at a single point					X
Multiple applications deployed on one server				X	
Application service dedicated on one server			X		
Deployment on any of the available resources					X
Independent deployment				X	
Service interface conformance		X			
Degradation (selected service denial, downgrade switches, etc.)	[RL]	X			
Repeated service call capability	[PTN]	X			
Third-party notification service	[TS]	X			
Asynchronous communication through message queue	[MQ]	X			
Service traffic warning	[MW-Monitor]	X			
System monitoring		X	X	X	

(*continued*)

Table 5-32. (*continued*)

Solution Metric	A-ESA	MS-C1	MS-C2	MS-C3	MS-C4
General fault-tolerance	[MW-Resilience]			X	
Resource isolation (CPU, threads, IO, etc.) using bulkhead, etc.		X			
Throttle (fail-fast, circuit breaker, flow control, timeout, failover, retry, etc.)		X			
Automatic failover including fault-tolerance and recovery		X	X		
Capacity resilience	[DB]	X	X	X	
Gray deployment service	[DEV]	X	X	X	
Direct deployment on the runtime platform		X	X		

Note:
– MS-C1: Core Services I MS-C2: Important Services I MS-C3: General Services I MS-C4: Utility Services

Microservice Pattern

A microservices architecture comes with technical challenges. Table 5-33 shows a list of patterns that address these challenges.

Table 5-33. *Microservice Pattern*

Category	Pattern	Action Description	A-ESA Notation
Isolation	Ambassador	Offload common client connectivity tasks	Middleware [MW]
	Backends for Frontends	Create separate backend services for different types of clients	Tech Service [TS]
Legacy Migration	Anti-corruption Layer	Implement a façade between new and legacy applications	Pattern View [PTN]
	Strangler Fig	Support incremental refactoring of an application	
Gateway	Gateway Aggregation	Aggregate requests to multiple individual microservices into a single request	Middleware [MW]
	Gateway Offloading	Enable each microservice to offload shared service functionality	
	Gateway Routing	Route requests to multiple microservices using a single endpoint	
	Messaging Bridge	Integrate disparate systems built with different messaging infrastructures	
Resilience	Bulkhead	Isolates critical resources	Middleware [MW]

Note:
– Source: https://learn.microsoft.com/en-us/azure/architecture/microservices/design.

Microservice Operational Architecture

Table 5-34 shows the microservice operational process using A-ESA notations.

Table 5-34. *Microservices Operational Architecture Process*

Step	Operational Process	Description	A-ESA
1	Choose a hosting option	Usually, two options as a tradeoff: – A service orchestrator that manages services running on dedicated nodes (VMs). – A serverless architecture using functions as a service (FaaS).	Key Choice [KC]
2	Decide on inter-service communication	Key choice decisions on the tradeoffs between asynchronous messaging and synchronous APIs, especially for distributed transaction handling.	Key Choice [KC]
3	Design microservice APIs	Meet the two different sets of API requirements: – Public APIs from client applications. – Backend APIs that are used for inter-service communication.	Service Interface [SI]
4	Specify API gateway	Determine the required API gateway capabilities through API gateway patterns such as Gateway Offloading.	Middleware [MW]
		Make architecture decisions about gateway technology.	Key Choice [KC]

Note:
– Gateway offloading, the process of moving functionality from individual services to the gateway, is a cross-cutting issue.

Style 11: EDM Architecture

An *event-driven microservices* (EDM) architecture that uses event streams as its flow is more than just a microservices architecture. By combining event-driven architecture (EDA) and microservices, an enterprise can build highly available, distributed, fault-tolerant systems that process large amounts of information in real time.

Table 5-35 shows the notations commonly used for modeling EDM architectures within the A-ESA scope.

Table 5-35. *EDM Architecture Notations*

EDM Terminology	A-ESA Notation
Event storming	Page Flow [PFV], Process [PRM]
Event requirements	Requirement/Rule [RQ]/[RL]
Problem space domain	Generic Service [GS]
Long-running microservices	Process [PS]
Domain/Subdomain	Domain [DM]
Partition	
Logical boundaries relating to a subdomain	
Bounded context	Bounded Context [BC]
Micro frontends	Frontend [FE], User Interface [UI]
Event format	Governance [GV]
Event schema (specification of data definition and triggering actions)	
Event policy	

(*continued*)

Table 5-35. (*continued*)

EDM Terminology	A-ESA Notation
Data liberation	Key Consideration [KC]
Eventual consistency handling decision	
Microservice tax/technical debt	Risk [RK]
Event consumer	App Logic Service [AS]
Application function	
Event producer	
Triggering logic	
Logic apps	
Event stream (as a service)	Data Service [DS]
Event sourcing/snapshots	
Buffering state	
Event store (event content)	
Event implementation	Tech Service [TS]
Event handler	
Event queue	
Customized queue	
Data contract	Service Interface [SI]
Protobuf (a well-known language-agnostic binary data format)	
Atomic function (FaaS function)	Service Component [SC]

(*continued*)

Table 5-35. (*continued*)

EDM Terminology	A-ESA Notation
Event (about the fact that happened)	Event Service [ES]
Event stream items (expressed in tabular or flow form)	Item [II]
Key/Value (a typical format representing events)	
Apache Kafka	Middleware [MW]
Event broker	
Debezium (a distributed platform product that changes DB info to event streams)	
Container Management System (CMS)	System [SY]
Event sink (generalization of components in an event streaming platform)	Platform [PL]
Dapr (a portal, serverless, event-driven runtime)	
External provider	Cloud [LO-CL]
External dependency	
Choreography design	Pattern [PTN]
EDM (Event-Driven Microservice) architecture	
Lightweight Microservices Framework	
CQRS (Command Query Responsibility Segregation)	
Business topology	Service Relationship [SRV]
Topology (graph-like structure)	

(*continued*)

Table 5-35. (*continued*)

EDM Terminology	A-ESA Notation
Trigger (incoming direction)	Directional Relationship [FW]
Action (outgoing direction)	
Event command (about the intent to do something)	
Event notification	
Cold/warm start	Property Attribute
Forward compatibility	
Stateful/stateless	

There are numerous alternative forms of event-driven architectures beyond EDM. For instance, Event-driven Middleware Architecture, another EDM, is a highly effective architecture capable of handling real-time events with exactly-once semantics, as evidenced by the Uber case, where the event-driven system is based on middleware (Apache Flink, Kafka, and Pinot).

Note that a clear model mapping will be helpful in the case of technical or integration architectures to see how they relate to the overall ESA.

HINT

What Is an Event? An event represents a change of state, a notable occurrence at a particular point in time. "Event" is another IT word that has multiple meanings. Event can be implied as a business event, an application event, an event-related rule, a technology event (event trigger, event handler, event queue, etc.), or a data service event. There are different constituents of event architecture, such as *event notification, event sourcing, CQRS,* and *event state transfer and data management.*

Style 12: Large-scale Website Architecture

This section explains a large website architectural style for high volume and high concurrency sites like Alibaba's online sales site. It is commonly referred to as the *high performance*, *high availability*, and *high throughput* architecture (the *3H architecture*).[13] It is a blend of architectural considerations from IT strategy, business context, application, data, and technology. Its key quality concerns are scalability, elasticity, availability, performance, and security, with scalability being the leading factor.

Scalability means that the architecture can continue to meet requirements as size and complexity change. In this way, a scalable architecture can meet flexible business needs. Scalability, flexibility, adaptability, and elasticity are synonyms that have slightly different meanings, but they are all about the ability to change.

Table 5-36 lists common rule-based scalability measures, while Table 5-37 lists core scalability measures from non-functional requirements: architectural *adaptability, availability, asynchronicity* and *analyzability*, an AAAA (quadruple A) architectural consideration in simple terms for easy remembrance. Note that adaptability is more about application architecture, availability about technical architecture, asynchronicity about data architecture, and analyzability about general architectural considerations.

---

[13] The reference measures for 3H architecture are response delay < 100ms, HA > 99.99%, and QPS > 100,000.

Table 5-36. *Rule-based Scalability Measures*

Principle	Architectural Consideration	A-ESA
Degrading	Degrading is enforced when there is a problem with the service or when it affects the performance of core processes such as the shopping cart service. Some scenarios cannot be solved by caching and downgrading, such as scarce resources (rush sales), writing services (orders), etc. Degradation levels include: – Business degradation: When high concurrent traffic strikes, ensuring that order placement and payment is part of core requirements. Some synchronous calls can be changed to asynchronous calls, high priority data or data with special characteristics can be processed first. – Degradable multi-level read service: For example, service calls are downgraded to read-only local cache – Page level: Partial page, degraded to other pages, dynamic to static page – Application service: Disable downstream application services – Business function: Disable inventory status – Data: Change remote service to local cache Implementation of degradation: – Manual and automated processes based on system throughput, response time, availability and other conditions – Use of middleware (such as Hystrix's circuit breaker) to implement downgrade	Requirement Rule [RQ-RL]

(continued)

Table 5-36. (*continued*)

Principle	Architectural Consideration	A-ESA
Throttle	Throttle solutions: – Dynamically adjust the threshold based on system throughput, response time, and availability. – Ultimately compromise service rather than no service. – Prevent malicious access: Malicious request traffic only accesses the cache. – Limit the maximum number of functional calls. For example, do not call this function more than once every 100ms. Throttle layers: – Application server layer: Load balancing, malicious request filtering, caching, request aggregation, A/B testing, etc. – Connection layer: Total concurrency limits (such as database connection pools, thread pools), MQ consumption rate, etc. – OS layer: CPU or memory utilization, etc. – Infrastructure layer: Network traffic tuning, etc. – Web server layer: Access by account, rule, IP, etc. – Application server layer: Sticky bits, etc. – Database layer: Critical DB zone, etc.	Requirement Rule [RQ-RL]
Separation	Separation is to implement hotspot isolation and prevent snowballing after a failure. Isolation layers: – Separation of UI and application logic – Isolation of hotspot business systems such as rush sales – Read and write isolation, e.g., replication for goods read, HA for goods write – Isolation of static page and dynamic pages – Separate provision of dynamic images – Process and resource isolation – Distributed data centers	Requirement Rule [RQ-RL]

Note:
– RQ-RL is a requirements specification, while GV provides details of governance functions and properties.

Table 5-37. *High-Availability Architectural Measures*

Metric	Activity	Architectural Consideration	A-ESA
Adaptability	Service-orientation	Service-based architectural design: – Service-based architecture or microservice architecture, independently deployable services for easy distribution and isolation – Inter-service communication frameworks such as Dubbo – Service control such as network traffic control restrictions/black and white lists	Functional services [MS]
	Process composability	Dynamic process invocation	Process [PS]
	State	State control: – Stateless service – Event control and state reversal (abort), etc.	Event Service [ES]
	Splitting	Multi-dimensional splitting by tier, module, read/write, static/dynamic, tech/utility vs. app/data, history data, etc.	Domain [DM]
		By business systems, functions, or other SLCs	Key Choice [KC]
	Rollback and version control	Rollback and version control: – Implement auditability, traceability, and rollback. If a program or data error occurs, you can roll back to the last correct version if a versioning mechanism is in place. – Codebase rollback – Static resource version rollback – Data version rollback – Deployment version rollback	Governance [GV]

(*continued*)

Table 5-37. (*continued*)

Metric	Activity	Architectural Consideration	A-ESA
Availability	Load balance	Load balancing and proxy service: – DNS load balancing – Multi-tier load balancing – Static load balancing – HTTP dynamic load balancing – Includes retry on failure, timeout, reverse proxy, proxy cache, keep-alive middleware, etc. – Using an appropriate load balancing algorithm: round-robin, ip_hash, hash key, etc.	Requirement Rule [RQ-RL]
	Scaling	Vertical scaling: – Application segmentation by business domain such as transaction and merchandise – Data partitioning by data domain such as merchandise and orders	Node [ND]
	Resiliency	Robust architecture: – Application system – Database – DNS routing policies	Middleware [MW]
	Health check	Detect single point failures in: Applications, middleware, DBs, nodes, network connections using various approaches such as active/passive, HTTP heartbeat.	Governance [GV]

(*continued*)

Table 5-37. (*continued*)

Metric	Activity	Architectural Consideration	A-ESA
	Virtualization	Resource utilization and allocation: — Virtual servers, virtual OS, containerization, etc. — Priority policies such as Level 1 systems use virtual application servers and Level 2 use VMs.	Virtual Service [VS]
	Backup	IT human resources	Role [RO]
		System nodes	Node [ND]
		Data	Data Store [DB]
		Running code	Deployment Package [DP]
	Timeout	— Business timeout and retry — Frontend timeout — Network timeout setting (retry delay, max retries)	Governance [GV]
	Security	Security attack prevention: — Web attack (SQL injection, XSS) prevention — Interface access authentication and authorization — Data encryption and integrity	Middleware [MW], Network [NW]

(*continued*)

Table 5-37. (*continued*)

Metric	Activity	Architectural Consideration	A-ESA
Asynchronicity	Concurrency	Event-driven architecture	Event Service [ES], Pattern [PTN]
		Queue technology with large traffic buffering	Message Queue [MQ]
		Connection pool with concurrent data fetching	Database [DB]
		Multi-threading, asynchronous programming, etc.	Governance [GV]
	Data splitting	Data distribution via message queues	Message Queue [MQ]
		Sub-database and sub-table, data sharding (e.g. by order ID)	DB Schema [SM]
		Data service heterogeneity: – Data heterogeneity (join queries in one DB) – Database heterogeneity such as Distributed File System (DFS), variety of relational and NoSQL DBs	Database [DB]
		Eventual consistency	Transaction Domain [DM-TR]
	Idempotence	Eliminate key and table duplication, such as repeated payments	Transaction [DM-TR]
	Caching	Caching architecture: – Frontend caching – browser, CDN caching – Reverse proxy – Multi-node caching – http servers, mirror servers, P2P, application server, distributed caching/Redis clustering, OS caching	Cache [CA], Operational [DEP]

(*continued*)

Table 5-37. (*continued*)

Metric	Activity	Architectural Consideration	A-ESA
Analyzability	Modeling	Architectural modeling: — Documentation of architectural governing ideas and significant architecture decisions — Division of capability responsibilities, domain/layer and service component management — Correlation between elements, views, and properties — Simulation for walkthrough and hotspot analysis — Ensure governance during implementation and deployment	Model Views [AOV], [VF], [ML]
	Pattern	Effective use of patterns such as: — Distributed inter-service communication — Clustering of deployment units — Layering (vertical, horizontal and cross-cutting) and partitioning — Asynchrony and concurrency — Caching and redundancy — Automation (deployment, version control, monitoring, etc.)	Pattern View [PTN]

(*continued*)

Table 5-37. (*continued*)

Metric	Activity	Architectural Consideration	A-ESA
	Emergency plan	Stress test: – Offline stress testers such as JMeter – Simulation stress test, tcpcopy and goreplay, which support offline and online recording and playback. – Actual stress test for significant cases such as web page, order processing, and fulfillment – Game day simulation with trusted seed data Capacity planning: – Estimate based on last year's traffic, historical records – Deployment rule of thumb: five times planned capacity – Capacity planning should be done before there are any predicted changes in usage patterns – Disaster recovery plan to ensure system availability through network switch, bandwidth routing , application system, DB	Deployment View [DEP]
	Monitoring	Tracking and monitoring: – Non-intrusive data collection, logging and alerting, distributed tracing – Monitoring of business metrics, security, application performance, servers, hardware, network – Configuration change health monitoring – Risk assessment based on the runtime patterns between SLAs (response time, throughput, etc.) and relationships between application services	Middleware [MW]

In short, for a 3H architecture, build with failure in mind. Always think about scaling and extreme burst cases.

Style 13: Agile and Digital Architecture

This section briefly describes key considerations in an Agile and digital environment where an *architecture-first* approach still plays a big role.

Agile Architecture

Agile architecture is about *Agile transformation*, which is a new way of managing product and project solutions. It pivots on improving the solution architecture and development process. Agile is one of the three broad lifecycle delivery approaches:[14]

- **Rapid:** Near-immediate implementation and fast release cadence such as old-fashioned rapid application development (RAD).

- **Agile:** Fast delivery cycles for specifically bounded functionalities.

- **Robust:** Risk and architecture-driven longer-term delivery of complex, large-scale enterprise solutions.

The depth and degree of adoption and customization of A-ESA in a lifecycle change approach depends on the scale, complexity, and environment of the solutions. Table 5-38 provides recommendations for reference adoption. Agile architecture has a number of important factors identified in the DORA reports: culture, product, technicality, and monitoring.

---

[14] *Enabling Enterprise Agility,* The Open Group.

Table 5-38. *Reference Adoption Recommendation*

Lifecycle Approaches	Adoption Recommendation
Rapid	Quick modeling views ([PFV], [AOV], [SRV])
Agile	Selected views with significant impacts ([DEV], [VF])
Robust	Full model views ([CAP], [PRM], [SIV])
Note: – Innovative solution is not necessarily a product; a small or fast solution. It can be complex coordination on a large scale that requires concerted effort.	

Overall, there are a number of architectural considerations for Agile architectures. Table 5-39 is a general mapping of these considerations to ESA.

Table 5-39. *A-ESA Mapping with Agile Terms*

Category	Agile Term	A-ESA Notation
Strategy	Strategic Theme	– Capability [CP]
	Principle	– Principle [PR]
	Resource	– Role [RO] – System [SY] – Data Service [DS] – Generic Service [GS]
	Simple Design	– Principle [PR]
	Minimum Viable Product (MVP)	– Deliverable [DL]
	Roadmap Stage	– View Frame [VF]
	Guardrail	– Governance [GV]

(continued)

Table 5-39. (*continued*)

Category	Agile Term	A-ESA Notation
Business	Value Stream	– Value Stream [VL]
	Bounded Context	– Bounded Context [BC] – Microservice [MS]
	Value Stream Mapping	– Value Stream [VL] – Capability [CP]
	Value	– Value [VL]
	Driver	– Principle [PR] – Requirement [RQ]
	Goal	– Principle [PR] – Requirement [RQ-GL]
	Alignment Diagram	– Architecture Outline [AOV]
People	Epic Owner	– Role [RO]
	Business Owner	– Role [RO]
	Lean Chief Engineer	– Role [RO]
	Product Manager	– Role [RO]
	Stakeholder	– Role [RO]

(*continued*)

Table 5-39. (*continued*)

Category	Agile Term	A-ESA Notation
Case Scenario	Journey Mapping	– Page Flow View [PFV] – Process View [PRM]
	Persona	– Role [RO] – System [SY] – Product [PD] – Generic Service [GS]
	Domain-Driven Design	– Domain [DM]
	Event Storming	– Process View [PRM] – Task [TK]
	Customer Experience	– User Interface [UI]
	User Story	– Use Case [UC] – Page Flow View [PFV]
	Epic	– Deliverable [DL] – Domain [DM]
Requirement	Feature	– Requirement [RQ]
	Modularity	– Non-functional Requirement [RQ-NFR]
	Integrality	– Non-functional Requirement [RQ-NFR]
	Evolvability	– Non-functional Requirement [RQ-NFR]
	Catchall	– Requirement [RQ]
Functional Service	Service	– Service Interface [SI] – Generic Service [GS]
	Application Function	– Application Interface [SI] – Service Component [SC]

(*continued*)

Table 5-39. (*continued*)

Category	Agile Term	A-ESA Notation
DevOps	Continuous Architecture	– DevOps [DEV]
	DevOps Culture	– DevOps [DEV] – Organization [ORG]
	Outcome	– Deliverable [DL]
	Architectural Runway	– Key Choice [KC] – View Frame [VF] – System Context [AOV]
	Automated Build	– DevOps [DEV]
	Continuous Deployment	– DevOps [DEV]
	CRC Cards[15]	– Principle [PR]
	Continuous Integration	– DevOps [DEV]
	Delivery Pipeline	– DevOps [DEV]
	Behavior Driven Development (BDD)	– Principle [PR] – DevOps [DEV]
	Acceptance Test Driven Development (ATDD)	– Principle [PR] – DevOps [DEV]
	Release [1]	– Deliverable [DL]
	Backlog [2]	– Requirement [RQ] – Risk [RK]

(*continued*)

---

[15] Class Responsibility Collaborator (CRC) cards are an object-oriented design technique that teams can use to discuss what a class should know and do and what other classes it interacts with should know and do.

Table 5-39. (*continued*)

Category	Agile Term	A-ESA Notation
	Technical Debt	– Requirement [RQ] – Key Choice [KC] – Risk [RK] – Governance [GV]
	Portfolio Canvas	– Value Stream [VL] – Use Case [UC] – DevOps [DEV]
	Product Increment [3]	– Deliverable [DL]
	Quality Properties	– Non-functional Requirement [RQ-NFR]
	Test-driven Development (TDD)	– Principle [PR] – DevOps [DEV]

Note:

– [1] Delivered by an Agile Release Train or a coordinated team of teams

– [2] Includes team backlog, program backlog, and solution backlog

– [3] The sprints in Scrum terms

– References

  – Agile Architecture Modeling Using the ArchiMate® Language

  – Open Agile Architecture™, a standard of The Open Group

Table 5-40 shows a simpler mapping based on the Agile architecture features.

Table 5-40. *A-ESA Mapping to Agile Architecture*

Agile Feature	Feature Item	Description	A-ESA Notation
Rapid Iteration	Rapid Prototyping	Fast visual rendering and page flow, based on the value proposition	[PFV], [VL]
	CI/CD	DevOps platform and automated tool support	[DEV], [MW]
	Spiral Model	Simple model of small iterations for MVP, each validated	[ML]
	Feedback Loops	Distributed decision-making	[KC]
User Experience	UI Design	User Experience (UX) Design for ease of use	[UI], [PFV]
	User-centric Test	Validation view from the user's perspective	[VLD]
	Performance	Application performance and responsiveness	[AP], [RQ]
	Documentation	Easy-to-follow user manual	[DF]
Automation	Coding Assistance	AI-powered code implementation	[MW], [CL]
	Test Automation	Test toolchain, test policy, and governance	[MW], [PR], [GV]
	Deployment Automation	Continuous integration and deployment	[DP]
	Version Control	Configuration MGT and version control for easy rollback	[MW]

Note:

– Enterprise Agile metrics imply efficiency, including percent of IT investment on run vs. grow, IT cost/value, value delivered (dollars or points), etc.

The Agile architecture is typically applicable to product deliverables or small solutions. For large solutions without proper modeling, Agile architecture presents some common issues that can be contained or mitigated with ESA, as shown in Table 5-41.

Table 5-41. *A-ESA Mapping to Address Agile Architecture Issues*

Agile Feature	Description	Common Issue	A-ESA Notation
Rapid iteration	Deliver functionality in small increments	Missing specification of significant quality attributes and constraints early in the solution	– Requirement [RQ] – Governance [GV]
User experience	Apply less formal user stories by epic or theme	Non-functional requirements not implemented to the same discrete extent as user stories describing quality attributes	– Use Case View [UCM]
Automation	Automate deployment and operation	Application-level run-time quality issues due to improper transition between functional services and operational services	– Deployment Package Mapping [DPM]

In Agile architecture, UX-focused[16] design is sometimes interpreted differently in real-world projects. For example, by definition, the Agile requirements topology is: *theme → epic → feature*, while in practice it may be *feature → epic → theme*. A product backlog item can be treated as an epic, feature, or user story that takes the form of *persona + need + purpose*. Therefore, a common understanding and mapping must be in place in the early stages of solution architecture. Table 5-42 is an example mapping. Note that any requirements that are not measurable are not valid inputs for ESA.

---

[16] UX: User experience

Table 5-42. A-ESA Case Scenario Requirement vs. Agile Input

Agile Term / Artifact	Theme	Epic	Feature	Story	Task
Use Case Model		X			
Requirement			X		
Use Case				X	
Process	X				
Task					X

Note:
- A theme is a broad area of focus that helps an Agile team to keep track of its organizational goals. Think of it as a label that can be used to group similar activities. A theme helps define the commonalities between different areas and unifies them under one heading.
- An epic is a large user story that cannot be delivered as defined within a single iteration.
- A feature refers to a specific functionality or service that satisfies a stakeholder's need. Each feature must outline the criteria involved and provide a specific business value.
- In consultation with the customer or product owner, the team divides the work to be done into functional increments called "user stories." A task is a subset of a story.

HINT

Agile vs. Lean: Agile focuses more on adaptability and rapid response to change, using an iterative and incremental approach that emphasizes team self-organization and customer collaboration. Whereas lean focuses more on efficiency and eliminating waste, emphasizing the continuous optimization and improvement of processes.

Theme vs. Task: Themes and tasks are similar in that they help categorize and bring a sense of order to the work management process. However, themes bring together a group of epics or initiatives together under a related banner. Tasks break down a story, creating subdivisions within an existing section. Themes are also broader in scope and can span the entire organization and relate to different epics, stories, and initiatives. Tasks, on the other hand, have a much more specific purpose. They are used only in the context of a single story and are not intended to be shared outside of that story.

Digital Architecture

In a digital transformation, an enterprise must adapt to emerging customer and business needs, in addition to business and IT alignment. Digital architecture can be a blend of intentional architecture and emergent architecture through continuous and evolutionary development. Digital architecture is essential to enable an Agile solution architecture, which is based on modular building blocks, to handle massive amounts of data. These building blocks can be configured from different perspectives (see Table 5-43).

Table 5-43. *Digital Architecture Perspectives and Building Blocks*

Perspective	Digital Intent	Model Representation	A-ESA Notation
Overall	Digital Strategy	Principle, Capability	[PR], [CAP]
	Corporate Brand Identity	Governance	[GV]
	Corporate Culture	Organization	[ORG]
Experience	Experience Design	Product Architecture	[AOV], [PD]
Work (Business) System	Journey Maps	Operations (Operating) Architecture	[PFV], [PRM], [UCM]
	Value Streams		[PRM], [CAP], [VL]
Technical System	Event Storming	Domain-Driven Design	Domain [DM], Use Case [UC], Process [PRM]
	Development Environment	DevOps	[DEV]
	Functional Building Blocks	Service Architecture	[SRV], [IRV]
	Technical Platform	Scalable Architecture	[DEP]

Agile and Digital Architecture

A simple difference between digital architecture and Agile architecture is that the former supports the new way of doing business while the latter supports the new way of doing solution delivery. Both require an Agile ESA to deliver quality results with increased efficiency. Table 5-44 demonstrates the differences between the two.

Table 5-44. Differences Between Agile and Digital Transformation

Area	Agile Transformation	Digital Transformation	A-ESA
Goals	Meet customer needs in a timely and effective manner while enhancing customer satisfaction.	Strive to boost the entire business.	Goal [GL]
Tasks	Drive specific tasks within the organization to facilitate focused change and localized improvement.	Adopt broad tasks for global improvement.	Task [TK]
Process	Constantly evaluate in search of incremental improvements.	Constantly look for new digital solutions and break free from technology debt.	Process [PS]
Note:			
– AI systems form an integral part of the process of digital transformation.			
– Reference source: https://www.danaconnect.com/digital-transformation-vs-agile-transformation-whats-the-difference/.			

From this comparison, you can see that Agile transformation and digital transformation have many similarities and their differences complement each other to form a more complete Agile and digital architecture.

HINT

MVP vs. MVA: Agile and digital architecture can use a design thinking methodology to gather requirements for case scenarios. Less formality can be given to the MVP (Minimal Viable Product) or MVA (Minimum Viable Architecture). If the MVP fails, there is no need for MVA. Otherwise, the MVA is essential to move the product solution forward.

Style 14: Cloud Architecture

In a public cloud environment, much or some of the architectural work is moved to the cloud provider. Many people thought that architectural issues would magically disappear after the workload was moved. However, there is still architectural work to be done during and after the cloud migration. Even a full-blown enterprise SaaS solution requires high-level enterprise solution architecture considerations, albeit little or no more functional design, operational runtime, and maintenance work.

Cloud Hosting Architecture

When hosting on a public cloud, the customer and cloud provider typically have a *shared responsibility model*, as outlined in Table 5-45. So, even for small IT projects or solutions based on cloud platforms, it is important to understand the architectural responsibilities.

Table 5-45. *A-ESA Mapping to Cloud Hosting Responsibilities*

Category	Provider	Customer	A-ESA Notation
Location and Zone	Data center location	Selection	– Location [LO]
	Edge locations	Selection	– Domain [DM]
	Regions	Selection	– Key Choice [KC]
	Availability zones	Platform access management	
System Infrastructure	Software infrastructure	Firewall configuration	– Middleware [MW]
	Hardware infrastructure	OS configuration	– Node [ND]
	Network infrastructure	Network configuration	– Network [NW]
	Security infrastructure	Security configuration	– Service Interface [SI]

(*continued*)

Table 5-45. (*continued*)

Category	Provider	Customer	A-ESA Notation
System Application		Application management	– Application [AP]
		Identify management	– User Interface [UI] – Middleware [MW]
		Data logic and encryption	– Data Service [DS]

Note:
- Application management [AP] may consist of microservice and/or various functional services.
- Also likely to be considered is the platform [PL] element for microservice DevOps runtimes such as Spring Cloud, Istio (Service Mesh), etc.

Cloud Architecture Migration

Cloud migration requires an *as-is* and *to-be* architecture assessment analysis and an architectural roadmap planning. The to-be architecture can be evaluated using many leading practices, techniques, and tools. Nevertheless, an ESA architect plays a distinct role in building a cloud architecture model. ESA can provide a clear architectural mapping between as-is and to-be, and see how the cloud works.

For cloud migration solutions, an assessment plan is first and foremost. ESA planning helps evaluate the migration strategy, compare cloud providers and products and make an informed architecture decision. Table 5-46 shows an example of cloud planning mapped to ESA.

Table 5-46. *Phase 1 Cloud Planning Mapping with ESA*

Planned Objective	Measurement Description	A-ESA Notation
Improved cost model	Shift from CAPEX to OPEX model	Value [VL]
Reduced risk	Use indicators of compromise (IOCs) to detect suspicious security activity, etc.	Risk [RK]
Data safety	Preserve critical information during cloud migration	Principle [PR]
Business continuity	Measure the availability of critical cloud migration services	Requirement [RQ]
Reduced time	Calculate how much time is needed to compete the migration	Requirement [RQ]
Improve operational efficiency	Develop operational measurements for hardware and network (disk performance, CPU utilization, etc.)	Non-functional Requirement [RQ-NFR]
User experience	Design user interface to meet the customer satisfaction scores for response time and error rate	User Interface [UI]
Migration decision	– Cloud migration decisions – Choose cloud type: private cloud, on-premises, replication, data center extension, public cloud, hybrid cloud, etc. – Select cloud provider and cloud migration technology, and avoid vendor lock-in.	Key Choice [KC]
Migration strategy	Decide on a cloud migration strategy: Rehost (lift and shift), Replatform, Refactor, Repurchase, Retain, or Retire.	Key Choice [KC]

Note:

– There are various tools for cloud discovery and assessment from several major vendors, including Microsoft (Azure Migrate Server Assessment Tool, etc.), Amazon (ASW Application Discovery, etc.), and Redhat (Redhat Migration Toolkits, etc.).

Cloud architecture planning is followed by cloud migration implementation. Table 5-47 is a list of architecture views (Phase 2) for cloud migration implementation. The views include all the elements as planned in Table 5-46. The two-phase approach better ensures the success of the cloud migration.

Table 5-47. *Phase 2 of Cloud Migration in A-ESA Views*

Implementation Activity	Migration Description	A-ESA Notation
Architecture Mapping	Clear architectural mapping between as-is and to-be	– Architecture Outline [AOV] – Service Interaction [SIV]
Application Refactoring	Software application migration	– Service Relationship [SRV]
	Significant refactoring cases	– Use Case Model [UCM] – Process View [PRM]
	Application dependency mapping	– Service Interaction [SIV]
	Cloud application migration patterns	– Pattern [PTN]
Development Environment	Application development and delivery environment, and automated pipeline	– DevOps [DEV] – Deployment Package Mapping [DPM]
Runtime Management	Environment security and operational security	– Architecture Outline [AOV]
	Monitoring, logging and alerting	– Operational [DEP]
Migration Risk Management	Phased migration, transition, rollback plan, and backup plan	– Architecture Overview [AOV] – Metrics View [MTS] – Validation View [VLD]

Cloud Service Selection

Specific cloud service providers, such as AWS, Azure, and GCP (Google Cloud Platform) can have a significant impact on the design and implementation of enterprise solution architectures, as well as interactions with operational and technical aspects. In addition to the key S-MAPS related factors, it is important to consider the features and services offered, the ecosystem and integration, the technical expertise and support, the innovation and market trends, the migration and portability, the localization and regional compliance, and so on, in order to meet your specific needs and business scenarios.

The cloud provider can influence architectural decisions, so the selection strategies often matter more than architectural efforts. Table 5-48 shows common cloud architecture choices about resource right-sizing and cost-saving options.

Table 5-48. *Common Consideration Choices of Cloud Architecture*

Option	Description	A-ESA
Demand Forecasting	Use data and forecasting tools to predict future demand and allocate resources more accurately.	[KC]
Cost Tools	Use cloud tools to monitor and manage costs.	
Rightsizing Resources	Check if resources are too big or too small and change them to match what you need.	
Multi-Cloud Strategy	Use multiple clouds to get the best prices and services from different providers.	
Serverless Architectures	Use serverless architectures like AWS Lambda or Azure Functions to cut costs by only paying for the compute time used.	
Shared Services	Consolidate resources and reduce redundancy with shared services.	

(continued)

Table 5-48. (*continued*)

Option	Description	A-ESA
Containerization	Use Kubernetes to manage resources and workloads.	
Cost-Effective Regions	Deploy resources in cheaper regions without losing performance.	
Reserved Instances	Buy discounted long-term instances for lower costs.	
Spot Instances	Use spot instances for non-critical, flexible workloads to save money.	
Optimized Storage	Use the right storage classes and services to cut costs, like archiving infrequently used data.	
Data Transfer Optimization	Cut data transfer costs by optimizing data transfer in and out of the cloud.	
Auto Scaling	Use auto-scaling to adjust resources based on demand, using them as efficiently as possible.	
Utilization Monitoring	Keep an eye on how you use cloud resources. If you don't need them, you can stop using them or make them smaller.	
Budgets and Alerts	Set budgets and alerts to track spending and avoid overspending.	
Lifecycle Management	Automate the management of cloud resources, ensuring that they are removed when no longer needed.	
Leverage Open Source	Use free, open-source software to cut costs.	
Software Licensing	Review software licensing costs and consider cloud-based options.	
Training and Awareness	Train staff on cost-effective practices and the importance of cost optimization.	
Regular Reviews	Review cloud spending and usage regularly to find ways to save money.	

Style 15: Security Architecture

Security is required in almost all aspects of architectural modeling. Because security is a cross-cutting concern, it is often considered as a separate architecture model profile that fits within the A-ESA model.

Security measurement can be derived from many standards and frameworks, including risk management, such as:

- ISO/IEC 27001:2013: Information Security Management

- Control Objectives for Information and related Technology (COBIT Framework)

- The Open Enterprise Security Architecture (O-ESA) standard

- The Open Information Security Management Maturity Model (O-ISM3) standard

- Sherwood Applied Business Security Architecture (SABSA)

- National Cybersecurity Frameworks

- ISO 31000:2009: Risk Management

Table 5-49 includes typical security measures and their associated A-ESA notations grouped by security category for solution architecture reference.

Table 5-49. *A-ESA Mapping to Information Security Practice*

Category \ Measurement	CP	SM	CB	CI	NW	SD	UE	SS
Asset MGT	X			X	X			
App control			X				X	
Availability			X	X	X	X	X	X
Business continuity		X	X	X	X		X	X
Cryptography		X	X				X	
Email							X	
Confidentiality		X	X				X	
Firewall					X			
Hand-held devices							X	X
Information security function		X				X		
Legal and regulatory compliance		X						
Local security coordination		X	X	X	X	X	X	X
Quality assurance						X		
Security architecture	X	X						
Security audit/review		X				X		

(continued)

Table 5-49. (*continued*)

Measurement \ Category	CP	SM	CB	CI	NW	SD	UE	SS
Security monitoring		X				X		X
Security testing						X		

Note:
- CP (Capability) maps to [CAP]
- SM (Security Management) maps to [CAP], [AOV], [PR], and [GV]
- CB (Critical Business Applications) maps to [AS], [TS], [DS], and [MS]
- CI (Computer Installation) maps to [DP] and [ND]
- NW (Networks) maps to [NW]
- SD (Systems Development) maps to [DEV] and [SRV]
- UE (User Environment) maps to [AP] and [UI]
- SS (System Software) maps to [SY] and [MW]
- The security practice is adapted from *The Standard of Good Practice for Information Security, ISF, 2007.*

In a cloud environment, security architecture addresses issues such as data encryption, access controls, and vulnerability management:

- Manage visibility across private, hybrid, and multi-cloud environments and enforce cloud security policies.

- Ensure security and governance, especially data security.

- Implement data encryption and data loss prevention (DLP), protect all applications, implement incident response plans, and meet compliance requirements.

- Use log management and continuous monitoring to leverage cloud detection and response methods.

- Protect containers and perform vulnerability assessment and remediation.

- Conduct penetration tests and perform intrusion detection, packet inspection, traffic analysis, and threat detection using a cloud-native firewall service or more advanced third-party tools.

You can map your own security architecture similar to Table 5-49, and the initial mapping is the foundation for the security model views.

Style 16: Reference Architecture

A reference architecture (RA) is a template solution architecture for a specific domain with an emphasis on commonality. It can be an *industry-specific reference architecture* (such as the retail reference architecture shown in Figure 2-13) or an *IT-centric reference architecture* (such as the data analytics reference architecture in Figure 2-14) and the *infrastructure reference architecture* (shown in Figure 2-15). The reference architecture can also be a *domain-based architecture* (such as the cell-based architecture shown in Figure 2-16) and the *baseline reference architecture* shown in Figure 2-18.

Table 5-50 provides a list of unique element notations common to four selected BizTech-style RAs. Each of these reference architectures can be rendered using its A-ESA element notations (check mark). See Figure 2-14 in Chapter 2 for an example model view of *Data Analytics RA*.

Table 5-50. *List of Reference Architecture Notation Representation*

Element Name	A-ESA Notation	IoT	Blockchain	Mobile	Service MGT
SRE Engineer	Role [RO]				X
Application Owner	Role [RO]				X
Offline Capabilities	Capability [CP]			X	
Process MGT	Process [PS]	X			
Process Operations	Process [PS]				X
Security	Requirement [RQ-NFR]	X	X	X	X
Information Governance	Governance [GV]	X	X		X
Visualization	User Interface [UI]			X	

(*continued*)

Table 5-50. (*continued*)

Element Name	A-ESA Notation	IoT	Blockchain	Mobile	Service MGT
Portal Service	User Interface [UI]		x		
Mobile Interface	User Interface [UI]			x	
Application	Application [AP]	x	x		x
Admin Service	Application [AP]		x		
Enterprise Application	Application [AP]	x	x	x	x
Mobile App	Application [AP]			x	
Application Logic	App Service [AS]	x			x
Membership	App Service [AS]		x		
Consensus	App Service [AS]		x		
Ledger	App Service [AS]		x		
Smart Contract	Service Interface [SI]		x		
Mobile Business Application	App Service [AS]			x	
Require Fulfillment	App Service [AS]				x
Incident MGT	App Service [AS]				x
Problem MGT	App Service [AS]				x
Change MGT	App Service [AS]				x
Configuration MGT	App Service [AS]				x
Analytics	Data Service [DS]	x			
Enterprise Data	Data Service [DS]	x	x	x	
Mobile Data Service	Data Service [DS]			x	

(*continued*)

Table 5-50. (continued)

Element Name	A-ESA Notation	IoT	Blockchain	Mobile	Service MGT
IoT Data Store	Data Store [DB]	x			
User Directory	Data Store [DB]	x	x	x	x
IoT Device Registry	Tech Service [TS]	x			
IoT Device MGT	Tech Service [TS]	x			
IoT Device Identity Service	Tech Service [TS]	x			
Transformation	Tech Service [TS]	x	x	x	x
Blockchain Integration	Middleware [MW]		x		
Blockchain Events	Event Service [ES]		x		
DevOps MGT	DevOps [DEV]				x
Edge Service	Edge Interface [EG]	x	x	x	x
Monitoring	Middleware [MW]	x	x		
API MGT	Middleware [MW]	x	x	x	
Blockchain Server Runtimes	Node [ND]		x		
IoT Device	Device [MD]	x			
Mobile Device	Device [MD]			x	
Connectivity	Network [NW]	x	x	x	x
Cloud	Cloud [CL]	x			

Note:

— Depending on the solution environment, element notation may vary. For example, the device registry could be a middleware element if a pre-built system software is used.

— Each reference architecture contains only significant and representative elements.

AI Reference Architecture and Pattern

Similar to the reference architectures (RAs) listed in Table 5-50, AI is a common RA that includes nine layers: 1) Business Insight Layer, 2) Data Layer (sources, collection, management, and storage), 3) Compute Layer (compute power either on-premises or via cloud services), 4) Model Layer (algorithm selection, feature engineering, model training, and evaluation), 5) Service Layer (API design, MSA), 6) Application Layer (UI, application logic), 7) Integration Layer (system integration, data exchange), 8) Security and Compliance Layer, and 9) Operations and Maintenance Layer (deployment strategy, monitoring, and logging).

AI requires considerations of all base architectures (BA, AA, DA, and TA). Table 5-51 shows an AI reference architecture mapping from a simplified A-ESA area perspective. See Figure 2-18 in Chapter 2 for an AI RA instance in the *model view* form.

Table 5-51. *AI Reference Architecture*

Area	AI Concept	A-ESA Notation
Case Scenario	AI usage requirement	Use Case [UC], Requirement [RQ]
	Machine learning/knowledge engineering case scenarios	Page Flow View [PFV]
	Algorithm specification	Rule [RQ-RL]
	AI workflow, business process, data stream, and app integration	Process View [PRM]
	Scoring	Task [TK], Item [II]
	Training	Task [TK]

(continued)

Table 5-51. (*continued*)

Area	AI Concept	A-ESA Notation
Outline	Machine learning reference architectures for Azure	Pattern View [PTN]
	AutoML (Automated Machine Learning), either supervised or unsupervised	Key Choice [KC], Task [TK], Middleware [MW]
	Pre-built AI models in AI Builder	Model [ML]
	Model selection	Key Choice [KC]
	Code languages (such as Mojo)	DevOps [DEV], Property Attribute
	RAG (Retrieval Augmented Generation)	Pattern [PTN]
Functional Service	Algorithm module	App Logic Service [AS]
	Deep learning/cognitive computing services	App Logic Service [AS], Data Service [DS], Tech Service [TS]
	Cloud cognitive services	Data Service [DS]
	Large context window service	Data Service [DS]
	Text to action transformation	Middleware [MW]
	Agent [1]	Adapter [EG], Middleware [MW], Tech Service [TS]
	Prebuilt AI	Service Interface [SI], Cloud [CL]
	Hyper parameter tuning	Service Interface [SI]

(*continued*)

Table 5-51. (*continued*)

Area	AI Concept	A-ESA Notation
Operational	AI technology stack — Storage capacity or volume — Networking and infrastructure — High computing, such as CPUs, GPUs,[17] FPGA,[18] and ASIC[19]	Operational View [DEP], Database [DB], Node [Node], Network [NW]
	LangChain/LangSmith	Middleware [MW], Platform [PL]
	Cloud-based bot service	Cloud [CL], Platform [PL], Middleware [MW]
	Apache Spark on Azure	Cloud [CL], Middleware [MW]
	Azure databricks runtime for machine learning	Cloud [CL], Middleware [MW]
	Autonomous systems, robotics, and bots	System Device [SY]

Note:

- [1] For AI solution architecture, AI terms need to be clarified and mapped to A-ESA. For example, the term "agent" can mean or include different things: system device, UI application, mobile assistant, data content connector, proxy, technical service, middleware, adapter, AI model or framework, pattern (such as REWOO, Self-Discover, Reflexion), etc. Only mapped model views can tell what a term really means in your AI solution.
- The microservice architecture style is a natural choice for AI's distributed networks.
- Reference source: https://learn.microsoft.com/en-us/azure/architecture/ai-ml/.

For small AI projects or applications, there may be more focus on using existing APIs and tools rather than mastering a full-scale reference architecture or building complex architectures. However, even for small or software projects, it is still beneficial to consider relevant architectural

---

[17] Graphics Processing Units

[18] Field Programmable Gate Arrays

[19] Application Specific Integrated Circuits

requirements (quality and constraints) and functional coordination. Table 5-52 shows A-ESA mapping with MVI (Model-View-Intent), a typical architectural pattern for AI solutions.

Table 5-52. *AI MVI Pattern*

MVI Layer	MVI Description	AI Component	A-ESA Notation
Intent	The system takes action or asks a question based on how it understands the user's input.	Intent recognition engines, NLU components, entity recognizers, and more	Value [VL], Requirement [RQ]
View	Examples of the data output include text, speech, and visuals.	Chat interfaces, voice interfaces, data visualization tools, and more	UI Service [UI]
Model	It's a machine learning model, knowledge base, or algorithm.	Neural networks, decision trees, support vector machines, etc.	Model [ML], Data Service [DS]
Controller	It collaborates models, views, and intents to work together.	Dialog managers, business logic processing units, I/O processors, etc.	App Logic Service [AS], Tech Service [TS]
Note: – NLU: Natural Language Understanding – The MVI pattern emphasizes the user intent, which is especially important when building interactive AI systems.			

Note that for many solutions, AI is just a learning system—a system that learns in the context of AI reference architecture. A large AI solution architecture could be a good fit with A-ESA. There are a few areas of the AI solution that may be worth focusing on, such as data policy and compliance, model selection, API layer access, data stream integration, computing power, and resources for cost effectiveness.

HINT

MVC vs. MVI: AI systems are often more complex than traditional MVC architectures in terms of processing logic. For example, an AI model may contain data processing, business logic, and decision-making processes at the same time, which are often separated in MVC architectures. MVI adds an Intent layer to better handle user input and system responses; however, it can also present some challenges, requiring more work to define and maintain the "intent" layer. In some cases, another pattern, MVU (Model-View-Update), which combines MVC and MVI, provides a clear way to build UI, and is better suited for apps with many states and interactions.

Architectural Evolution

Unlike a *building* architecture, IT architecture experiences constant changes. A large ESA is often a mixture of different architectural styles. An ESA may also be a form of minimal architectural design based on an existing architectural framework and platform. A big, up-front architecture or one-time deal architecture without much thought is not what A-ESA is intended for. A successful IT architecture can only grow incrementally from simple to complex constructs. Many enterprise solution architecture cases have proven this time and time again. Table 5-53 shows a typical *evolutionary* process for large enterprise solution architectures.

Table 5-53. *Architectural Styles for Different Scale and Complexity*

Scale	Arch Style	Description
Low	Monolithic Architecture	Typical three-tier architecture: presentation layer, application service layer, and database layer.
Low-Medium	Monolithic Architecture with scalability	Enhanced multi-tier architecture with vertical and horizontal scaling, load balancing, active/passive or active/active pair high availability.
Medium	Monolithic Architecture with modular applications and more scalability	Application refactoring or resign, more granular application services with remote procedure calls or service interface invocations, supported with greater system scalability.
Medium-High	Service-oriented architecture (SOA)	Flexible architecture with GUI integration, business process management, application service distribution, enterprise service bus, data federation and aggregation, using the teamwork development, and SOA framework such as Dubbo.
High	Microservices Architecture (MSA)	Featuring independently deployable services, DevOps, distributed databases, and a scalable and resilient platform.

HINT

Buzzword Architecture: Buzzword-driven architecture is often used pejoratively to describe the superficial adoption of architectural terms, while evolutionary architecture is a thoughtful approach to designing systems that can evolve and adapt to changing needs. Some IT folks tend to jump on the IT bandwagon or rush to architect solutions. The resulting architecture debt is often a big waste, or likely a slow burn.

Summary

As you can see from this chapter, each architectural style provides a unique guidance for business alignment, technology selection, NFR consideration, architectural or development efficiency, team coordination, reusability, adaptability, deployability, and innovation. You may have noticed that, as in reality, some of the architectural styles overlap to a certain degree.

Architectural style allows for modeling in a specific area with a correct set of elements. You can refer to the architectural styles in this chapter to create models that fit your solution environment. Note that the architectural styles in this chapter are *in summary form*, with brief descriptions and tabular form mapping. You must create model views to make them vivid and solution-specific. To elaborate further, each style could have been more thoroughly delineated with the aid of model view examples. Refer to some of the styles that were partially demonstrated in Chapter 2.

If you are unfamiliar with the style, you may find it difficult to understand the *condensed* narratives at first glance without visual views or detailed examples. However, when you put it into modeling practice, you will see what it means in ESA. Practice and let the results speak for themselves.

With increased automation, augmented AI, changing platforms, and technologies, new architectural styles will emerge. However, they are all based on the basic architectural styles. This chapter has shown that different chosen architectural styles have A-ESA notations in common, which reduces the complexity of multilingual styles.

Each style will be slightly different when applied to each solution case. However, the architectural styles as a whole follow a common set of architectural practices and fitness rules. The next chapter shows you selective governance and architectural techniques to support these styles in terms of their underlying architectures.

CHAPTER 6

A-ESA Governance Techniques

Only a model with governance has the power of conformance.

Architectural experience, whether personal or from others, can be systematized into techniques. This chapter touches on techniques, more on the art of IT architecture. These techniques, or simply *guides*, aim at governance, which is a set of rules and policies based on leading practices and lessons learned to achieve architectural goals. Many of the techniques in this chapter are summarized from practical cases. Figure 6-1 shows this chapter's holistic view of enterprise IT solution governance techniques, along with solution management context.

© Sean (Chunhong) Gu 2024
S. Gu, *Mastering Enterprise Solution Modeling*,
https://doi.org/10.1007/979-8-8688-0992-7_6

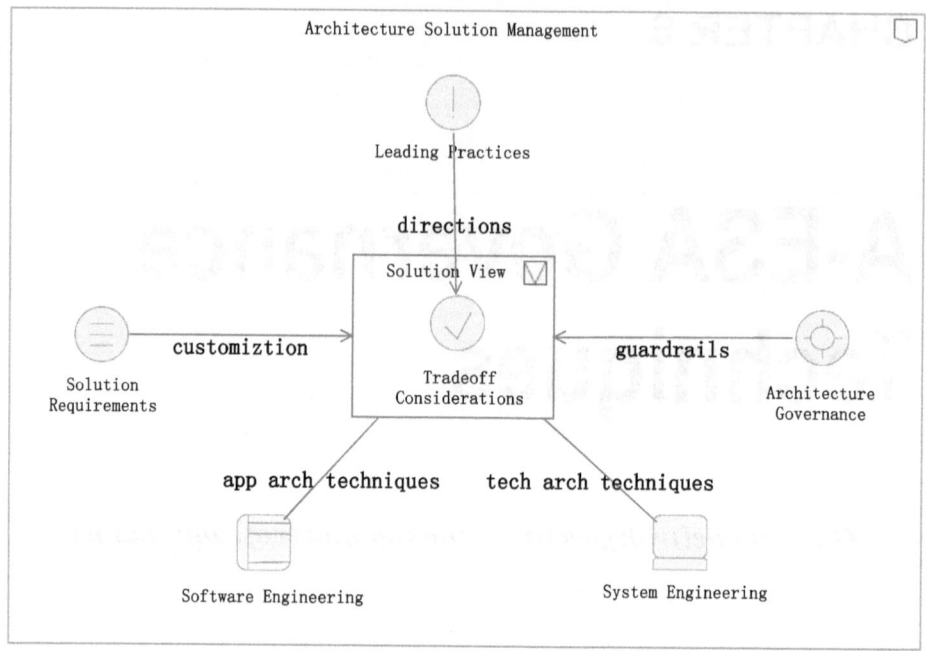

Figure 6-1. *Theme of governance techniques*

Guide 1: Customizing Modeling for Purpose

Since each solution has its own requirements, A-ESA requires customized adoption. A-ESA stands for Agile modeling and advocates effectiveness over formality. To use the A-ESA model effectively, you need to familiarize yourself with the A-ESA model specifications and tools and tailor modeling to your needs. Follow the model guidance in this book and refactor the model through iterations.

Adopting A-ESA Model Form(s)

The term "model" can refer to several different things. The Agile ESA model is designed to take these forms:

- **Correlated View Model:** The standard form of the A-ESA model that includes a complete set of correlated views and elements with an S3 (simplicity, significance, and systematics) focus.

- **Sketching** (the least formal model views): Quick diagrams without model-level element correlation.

- **Lean Model**: The simplest set of views and elements for initial or quick modeling.

- **Architectural PoC**: Proof of concept for an ESA without property specification details.

- **Architectural Walkthrough**: Significant case scenario simulation, hotspot walkthrough, or troubleshooting diagnosis with property specifications and tooling aid.

- **Lightweight EA Model**: Including strategic capability model, business model, economic model, and so on, in a compact form.

- **Architectural Compliance:** A top-down reference architecture, governance architecture, or templates as a repository for IT methodologies.

- **Profiled Model:** Serving as a major consideration of important profile(s) or architectural interoperability.

- **Solution Design Guidance** (the most detailed model specification): Modeling with governance specifications for an end-to-end implementation using assistive elements.

- **Digital Twin Model**: The full A-ESA model that reflects the running solution system as a holistic modeling with much automation and AI support (note that it must be presented in a clear and abstract form to be meaningful for architectural thinking).

A-ESA can take many forms, but the architect must be clear about its intent and effect. A-ESA, as an IT architecture language, can be represented or elaborated in an easily understandable form via PPT, DOC, or images. However, to be architecturally meaningful and maintainable, an A-ESA model must be kept at the ESA abstraction level, correlated, and consistent with the elements of the base model.

HINT

Modeling from a Developer's Perspective: You can use a software solution model such as C4, or more simply, many software visual design tools that support layered views, context diagrams, dependency diagrams, and code maps. Or you can simply do reverse engineering to automatically generate views for analysis and refactoring. Certainly, the logical views generated by the visual design tools will enhance architectural thinking and modeling needs. Or you can use software visual design tools to specify custom domain-specific language (DSL) as part of architectural governances.

Following a Modeling Process

A-ESA is an iterative and incremental development process. Its full architecture views and elements inherently contain all architecture processes. A-ESA typically includes the following:

- Help identify gaps, objectives, solution environment, migration or integration strategy, team and stakeholder RACI, deliverable breakdown structure or milestones (if solution management is part of it), and so on.

- Adapt or customize ESA model specification and solution architecture styles.

- Perform requirement mapping, especially significant case scenarios.

- Create governing views such as system context, building blocks (capability and value stream mapping), reference architecture/patterns, and so on.

- Perform metrics mapping for key architectural considerations

- Create logical, functional, and operational models.

- Iteratively validate the solution(s) and enforce architectural governance.

A-ESA does not impose any architectural process specifications. You can either follow the inherent A-ESA iterative modeling process, the informal A-ESA SMVG cycle, or adopt a process you are familiar with, such as ADM (architecture development method) or the well-known PDCA (plan, do, check, and act), and so on.

It's important to remember that a well-defined architectural process is only as good as the modeling process that goes with it.

Customizing A-ESA Architectural Style

You can use different architectural styles to suit the needs of each solution. A solution architecture is typically a mixture of different architectural styles. In addition, a focused and customized architectural style is often very helpful in addressing a unique, significant issue. For example, a caching architecture may be an architectural style that specializes in tradeoffs between business logic consistency and performance improvement; an API contract architecture may clarify key issues for inter-service communication. Since no single significant issue stands alone, the holistic modeling approach is often the way to go.

Customizing A-ESA Solution Operating Model

A-ESA's Solution Operating Model (see Figure 3-2 in Chapter 3) reflects the solution team organization and therefore the architectural style adopted.

In a traditional functional organization, the development team and the operations team are separated as *software engineering* and *systems engineering* and are responsible for functional services and operational/platform services, respectively.

In a DevOps environment, the development team is responsible for the functional and operational services. Typically, the infrastructure setup or configuration is done by a dedicated platform team.

In some organizations, the data information level is handled by the application team, so the Solution Operating Model presents only the application and technical levels.

In a cloud platform environment, the operating model is divided into the upper application level (application, FaaS, etc.) and the lower platform level (PaaS, CaaS,[1] IaaS, etc.). PaaS can be divided into aPaaS (application level) and iPaaS (infrastructure level).

Therefore, the Solution Operating Model will look different depending on each organizational structure, development culture and environment, and architectural preference. Each ESA must have a clear operating model that aligns well with the organizational structure and architectural intent.

Some organizations apply multi-layer conceptual, logical, and physical level model views to the operating model, which appears more systematic, detailed, and complete. For ESA, however, the complex model has proven time and again to be counterproductive and less useful.

Customizing A-ESA Areas/Views

The lead architect for the enterprise solution is responsible for the adoption decisions of the model areas. Although it is recommended to use the six areas defined in the A-ESA approach (see Chapter 1), variations may be more applicable to your solution environment. The views can be grouped according to your team structure, job role responsibilities, and stakeholder viewpoints.

For example, you can simply define three architectural areas (see Table 6-1) that show the focus of each view in terms of enterprise consideration, solution overview, and solution services.

---

[1] Container as a Service

Table 6-1. Views by a Broader Area Scope

Area Scope	Architecture View
Enterprise Consideration	— *Capability (Organization View)* — Architecture Metrics — *Enterprise DevOps*
Solution Overview	— *Pattern (Reference Architecture View)* — Introductory Outline (Block View) — Architecture Validation
Solution Services	— Case Scenario (Use Case View, Process View, Page Flow View) — Functional Service (Relationship and Collaboration View) — Operational Service (Package, Deployment View)
Note: — The italics are optional depending on the architectural style and solution requirements.	

As practically used in project settings, you can adopt a nine-grid approach (see Table 6-2) or a detailed approach (see Table 6-3) to partition the views in a more elaborate way.

Table 6-2. Nine-grid Architecture Areas

CAPABILITY	CASE SCENARIO	ARCHITECTURE PATTERN
OVERVIEW	FUNCTIONAL SERVICE	METRICS
DEVOPS	OPERATIONAL SERVICE	VALIDATION

Table 6-3. *Model View Variations Based on the Situational Needs*

Area	Arch Planning	Enterprise Solution Arch	Governance
Enterprise Solution Analysis			
Enterprise	– Strategic roadmap – High-level capability – Organization – Business architecture – Value stream	– Capability	Architectural conformance such as IT standards
Architecture Outline	– As-is environment – System context – Solution architecture overview – Solution definition – Assessment and estimation	– Architecture overview – Domain definition (tier, layer, domain, platform, environment) – Application architecture – Data architecture – Technical architecture – Integration – Migration/transition if any	
DevOps	– DevOps (organization, culture, process)	– DevOps (process, platform/tools) – Test cases	

(*continued*)

Table 6-3. (*continued*)

Area	Arch Planning	Enterprise Solution Arch	Governance
Solution Architecture			
Case Scenario	– User story theme – Business process – Page flow	– Use case – Application process	User support guidelines
Functional View		– Functional service collaboration/interaction – Functional service relationship – Information service relationship – Service component realization	Viability assessment
Operational View		– Deployment package mapping – Operational	
Solution Governance			
Metrics	– Metrics focusing on principle and risk – Requirement mapping	– Metrics focusing on key considerations	Metrics on governance and risk
Pattern	– Reference architecture	– Pattern	Governance functions
Validation	– Testing plan	– Relationship cross-checking and validation	Solution asset

Note:

– Some of the variations (such as sketch views) are not directly correlated to the model. They serve as conceptual diagrams for model mapping.

– A smaller set of views may be chosen if the A-ESA model is for a specific solution purpose or a narrow scope.

You can also simply focus on one solution area, such as *Green IT* architecture, and give a full consideration from its affected or highlighted views, such as software optimization, virtualization, and so on.

Although A-ESA has a flexible definition of *area* and *view*, the architectural coverage must be complete or well balanced, specifying all required views, element sets, and significant case scenarios. This is enforced and enhanced by the A-ESA functional, operational, and metric elements, which support structural and decisional architectures.

Optionally, depending on your architectural intent, you may create a set of *conceptual views* that do not require element correlation as an initial architectural assessment or sketch presentation. You can create a set of *logical model views* that have no association with physical elements. You can create a set of *physical model views* that contain real middleware, products, tools, and so on, for a one-shot model that is easy to understand. Or you can create a set of *mixed-style model views* for different stakeholders and viewpoints.

Whatever you choose, keep it simple. Use lean mode views whenever you can.

Customizing A-ESA Elements

The key to A-ESA is the choice of element notations, which should balance architectural clarity, visual expressiveness, and simplicity.

For very large and complex enterprises that include interrelated enterprises, a full model integration will apply. For a medium enterprise, a boilerplate model with less complexity and property specifications is appropriate. For an Agile or small solution, the simple set of *lean mode* elements is recommended. See Table 3-7 in Chapter 3 for different element model modes.

Here is the general process of element adaptation when not using lean mode:

- Choose from A-ESA basic or standard elements.

- Determine assistive elements, if any, and/or define your own subcategory elements.

- Customize chosen elements with proper naming and imaging.

For the same enterprise solution, an ESA architect may choose a different set of elements for effective modeling that fits the solution environment. Table 6-4 shows a simpler, customized set of A-ESA notations for the EDM architecture (see Table 5-35 in Chapter 5) in an Agile approach.

Table 6-4. *Lean Set of A-ESA Notations for EDM*

Area	A-ESA Notation	Prefix	Example EDM Element
Case Scenario	Domain	DM	Business topology
			Problem space domain
			Logical boundaries relating to a subdomain
	Requirement/Rule	RQ/RL	Event requirements
Metrics	Key Choice	KC	Microservice tax/technical debt
			Event format
			Event schema
			Event policy
			Data liberation

(continued)

Table 6-4. (*continued*)

Area	A-ESA Notation	Prefix	Example EDM Element
Functional Service	Service	GS	Event consumer
			Event producer
			Triggering logic
			Atomic function, FaaS function
			Data contract
			Long-running microservices
	Event	ES	Event stream
			Event service
			Event handler
	Item	II	Event stream items
			Key/value
	Database	DB	Event data
			Event store
Operational Service	Event Queue	EQ	Event queue
			Customized queue
	Middleware	MW	Debezium
			Apache Kafka
			Event broker

You may want to consider creating an ESA architecture using two different sets of A-ESA elements, as seen from Table 5-35 and Table 6-4. It may be the case that the simpler set is a better fit.

In general, *don't try to use more element types of ESAs for modeling.* The important part of A-ESA modeling is to specify and manage elements that share the same SLCs, indicated by the same prefix in the A-ESA model. Using more types of elements will make systems thinking less effective and lead to modeling complexity. However, for visual effect, different images or notation shapes can be used to distinguish one element from another within the same SLC group.

Guide 2: Applying Leading Practice

There is no single best architectural solution, but there are many leading practices: reference architectures, patterns, applicable architectural styles, proven expertise, architectural axioms, and so on.

Architectural Tenet

Architectural axioms, laws, and theorems are golden rules to follow in most cases. Tables 6-5 and 6-6 exemplify architectural axioms and laws that speak to how good IT architectural modeling work should be.

Table 6-5. *Selected Axioms and their A-ESA Implications*

From Whom	Axiom/Law Quotation	A-ESA Implication
Antoine de Saint-Exupery	"Perfection is achieved, not when there is nothing more to add, but when there is nothing left to take away."	Model simplicity [Chapter 1]
Greg Nyberg	"'Perfect' is the enemy of 'good enough'."	Modeling significance [Chapter 2]
Erik Doernenburg	"Get the 1,000-foot view", "this 1,000-foot view would provide information at the right level."	Importance of the architectural thinking framework [Chapter 3]
Grady Booch	"Architecture represents the significant design decisions that shape the form and function of a system, where significance is measured by the cost of change."	Architectural measurement [Chapter 4]
Randy Stafford	"There is no one-size-fits-all solution."	Architectural style [Chapter 5]
Mark Ramm	"Chances are, your biggest problem isn't technical."	Architecture governance and techniques for right solution modeling [Chapter 6]

Table 6-6. Examples of Architecture Solution Laws

From Whom	Statement
John Gall	Gall's Law: "A complex system that works is invariably found to have evolved from a simple system that worked."
Mel Conway	Conway's law: "The structure of a system is determined by the communication patterns of the people who design it."

There are architectural theorems that we must adhere to. The well-known CAP theorem, for example, is a must in enterprise solution architecture. It states that any distributed data store can only provide two of the following three guarantees[2]:

- **Consistency:** Every read receives the last write or an error.

- **Availability:** Every request receives a (non-error) response, with no guarantee that it contains the latest write.

- **Partition tolerance:** The system will continue to operate in the event of any number of messages being dropped or delayed by the network between nodes.

HINT

Counterexample: There are exception cases, for example, sometimes the organization is shaped by the structure of the system. There are also workarounds, for example, CAP balancing can be bypassed and becomes less restrictive with the advent of distributed transactional architecture (see Table 5-25), and if you don't model your entire system within a CAP, you can use a dynamic CAP, where different parts can have different CAP positions.

---

[2] Also known as "The Impossible Trinity."

Architectural Pattern and Template

A reusable solution architecture pattern can be applied to different solution requirements, while a template is a pre-designed layout that serves as a starting point for a specific solution.

Solution Pattern

A good architectural realization certainly depends on many solution patterns that contain proven solution implementations and provide consistency with typical properties of similar things or connections. An ESA architect needs to be aware of the enterprise solution patterns, especially the integration and distributed patterns, as illustrated in Table 6-7.

***Table 6-7.** Integration and Distributed Patterns*

Pattern Name	Usage Intent	A-ESA Notation
Sharded Services	Increase the size of the data being stored in the service.	– Database [DB] – Data Service [DS]
Functions and Event-Driven Processing (FaaS)	Strongly decouple each piece of the service.	– Cloud [CL] – Event Service [ES]
Ownership Election	Handle concurrent data manipulation.	– Data Service [DS]
Coordinated Batch Processing	Pull multiple outputs back together for some sort of aggregate output.	– Process [PS] – Event Service [ES]
Idempotent Receiver	Safely receive the same message multiple times.	– Message [MG] – Data Service [DS] – Event Service [ES]
Circuit Breaker	Handle faults that might take a variable amount of time to recover.	– Middleware [MW] – Tech Service [TS]

Note:
– Recommended solution architecture pattern books include "Enterprise Integration Patterns : Designing, Building, and Deploying Messaging Solutions" by Gregor Hohpe, Bobby Woolf, et al. and "Designing Distributed Systems: Patterns and Paradigms for Scalable, Reliable Services," by Burns Brendan.

Understanding the architectural pattern is one thing, applying it is another. Presenting a pattern in a model view will help clarify a pattern that fits the solution. As you can see from some of the solution pattern views in Chapter 2, an architectural solution pattern usually has its own variations.

Service Template

The solution pattern includes pattern frameworks and templates. A key to successful solution delivery is to specify well defined service templates that cover both *generic* and *specific* service specifications. Table 6-8 shows a service realization template using a chassis pattern. As can be seen, a service often requires mapping to different A-ESA notations.

Table 6-8. *Service Realization Template Using a Chassis Pattern*

Service Template	Template Item	Generic	Specific	A-ESA Notation
Functional Service	Service skeleton	X		Document File [DF]
	User interface, App logic, data, or tech service		X	UI [UI], App Logic [AS], Data [DS], Tech [TS]
	Service interface		X	Service Interface [SI]
Cross-cutting Attributes	Security (e.g., access token)	X	X	DevOps [DEV]
	Logging (e.g., log4j)	X		
	Health check (e.g., monitoring URL)	X	X	
	Test support	X		
	NFR metrics (response time, throughput, etc.)		X	Requirement [RQ-NFR], System [SY]
	Persistency (transaction mechanism)	X	X	Pattern [PTN], Data Service [DS], Tech Service [TS]

(continued)

Table 6-8. (continued)

Service Template	Template Item	Generic	Specific	A-ESA Notation
Build Logic	Configuration (db, mq, etc.)	x	x	Virtual Service [VS], Tech Service [TS]
	Dependencies		x	
	Development tool access	x		Tech Service [TS],

Note:

— In the development environment, use tools to glue generic and specific templates together to form a cohesive service.

Guide 3: Key Tradeoff Consideration

All significant considerations in ESA are tradeoffs. For each ESA, there are many overlapping non-functional requirements. Understanding the representative quality attributes is a prerequisite for rational tradeoff balancing.

Similarity and Dissimilarity Analyses

This section describes similarity analysis and dissimilarity (tradeoff) analysis.

Similarity Analysis

One issue is the confusion often seen in *closely paired architectural terms* such as business service vs. application service, business process vs. application process, domain service vs. application service, domain vs. module, application service vs. functional service, service component vs. component, technical partition vs. domain partition, problem domain vs. solution domain, and so on.

Similarly, many non-functional requirements are mutually influencing *overlapping NFR pairs* such as reliability vs. availability, scalability vs. high availability, changeability vs. agility, composability vs. modularity.

Another issue is the *implicit NFRs*. Some NFRs are actually boiled down from business and functional aspects, such as business goals, objectives, values, or general requirements. For example, the statement "service components shall be loosely coupled" may mean maintainability, or more specifically, adaptability. Sustainability can mean covering availability, reliability, testability, and maintainability.

There is a similar problem with the *vague NFR terms*, which are often used loosely in the solution project environment. For example, maintenance can mean a large NFR coverage, or it may mean only one or a few specific NFRs.

An ESA solution architect must ensure that the architecture element or view represents the specified NFRs. In particular, the vague terms such as maintainability need to be clearly specified.

Table 6-9 illustrates the *implications* and similarities between some commonly used NFR terms.

Table 6-9. *Clarification of NFR Vague Terms*

	Maintainability	Efficiency	Reliability	Complexity
ADAPTABILITY	X			X
ANALYZABILITY	X			X
AVAILABILITY			X	
CAPACITY		X		
MATURITY			X	
MODIFIABILITY	X			X
MODULARITY	X			
PERFORMANCE		X		
RECOVERABILITY			X	
REPLACEABILITY	X			
REUSABILITY	X	X		
ROBUSTNESS			X	
TESTABILITY	X			X

A simple approach to eliminating or clarifying *weasel words* is to have a clear comparison of synonym definitions, as illustrated in Table 6-10. See also Table 5-16 in Chapter 5 for fuzzy business term mapping with NFRs in the business architecture.

Table 6-10. *Flexibility vs. Adaptability*

NFR	Interpretation
Flexibility	An external factor initiates a change in a system, resulting in a transition from one state to another.
Adaptability	An internal impetus for change is the driving force behind the transformation of a system within a given state.

A good way to distinguish between pairs is to specify their corresponding notations in an A-ESA model. For example, the *business process* in a scenario case view is represented by *role* and *task* elements, while the *application process* in a functional service view is represented by *functional services*.

The general advice is to use fewer but the most representative NFRs and make sure that all stakeholders are on the same page. The art of tradeoffs is to avoid pitting two conflicting non-functional requirements against each other, and to find a *mutually beneficial balance*.

HINT

BizTech Communication: Vague business terms such as flexibility and speed need to be translated into measurable NFRs, and vice versa. All stakeholders need to agree on such a translation based on the architectural interpretation. Use modeling as a BizTech language in the form of a cartoon-like view, analogy, metaphor, matrix view, or visual cross-mapping for mutual understanding.

Tradeoff Analysis

Architectural modeling requires tradeoffs between competing forces. It involves important, complex decisions that are often difficult to make, such as resilience vs. performance, transactional integrity vs. availability, modularity vs. integrality, and agility vs. safety.

In a sense, the tradeoffs can be simply associated with the five measurement criteria (S-MAPS)—sustaining cost optimization, manageability, availability, performance, and security—as mentioned in Chapter 4. All tradeoff mappings are ultimately related to cost effectiveness. As shown in Table 6-11, these tradeoffs are generally iterative in this order:

1. Tradeoffs between performance and availability (the system behavior concern)

2. Tradeoffs between security and the whole solution (the compliance concern)

3. Tradeoffs between manageability and all runtime attributes (the non-runtime concern)

4. Tradeoffs between economic cost and overall solution (the economy concern)

Table 6-11. *Common NFR Tradeoff Steps in ESA*

System Behavior	SLC PLUS	SLC Minus
Performance vs. Availability		
Horizontal load balance	High availability	Increased latency in spanning nodes
Data partition or sharding	Better scalability	Increased latency and complexity
Data offloading to caches	Better performance	Reliability of invalidation
Auto-scaling	Spike or burst prevention	Reduced efficiency due to idle capacity
Horizontal load balance	Performance efficiency	Less scalability

(*continued*)

Table 6-11. (*continued*)

System Behavior	SLC PLUS	SLC Minus
Security vs. Solution Quality		
High level security control	Better compliance	Increased implementation and operational complexity, reduced performance
Strict identities verification	Better compliance	Increased dependencies
Colocation	Access isolation	Difficult to implement role-based security
Manageability vs. Runtime Quality Attributes		
High composability	Better maintainability and	Increased effort and skill requirements, reduced performance
High deployment frequency	Risk mitigation	Less system stability
High observability	Increased transparency	Likely data security leaks
Better operational management	Higher productivity	Increased resource spending
Sustaining Cost Optimization vs. Solution Quality		
Cost-savings on redundancy	Less costly and fewer efforts	Reduced scalability and resilience
Workload resources consolidation	Cost optimization	Increased likelihood of failure

Note:

— Runtime quality NFRs are availability, performance, and security.

— Solution quality NFRs are manageability, availability, performance, and security.

There will be a long list of such comparisons as in Table 6-11. Many are common sense to an experienced ESA architect or IT guy, but they need to be embodied in the model.

In a solution environment, many comparisons are typically made between different middleware products, patterns, architectural styles, and solution approaches. A-ESA's metrics view plays a major role in correlating and capturing the considerations. For example, when choosing between an orchestrator and a serverless approach, you can use these primary metrics criteria: manageability *principle*, application integration *requirement*, and economic *constraint*.

HINT

Tradeoff Analysis Record: You can follow the *architecture decision record (ADR)* template[3] of your choice or use the *architecture tradeoff analysis method (ATAM)* to balance competing priorities. But the analytical formality is less important than the *model thinking*, of which tradeoff analysis is an integral part. Examples of tradeoffs can be found in the solution profiles in Chapter 2.

Key Architectural Considerations

Architecture modeling is about more than simple tradeoffs. It is also about architectural significance and business impact. For modeling purposes, only considerations that are *significant and tough choice*s need to be specified in the model views. In A-ESA, the decision element is termed as a key consideration, and it's more about the key or big choice than the general architecture decision. A key architectural choice can sometimes achieve a sustainability objective with little effort by making clever tradeoffs.

---

[3] It usually includes rationale, solution context, assumption, implication, relevance, alternatives, pros and cons, decision, consequence, responsibility, status, timestamp, and so on.

The key choice decision does not necessarily come from one architect, but likely from *distributed decision-making.*

Big Architecture Decisions

If you have many decisions to make, stick to the big decision(s), such as shared or distributed database. One big decision can offset many small decisions. So, choosing sensible modeling *entry points* can be halfway to a successful solution.

Here is an example of a key consideration: the choice of a distributed database selection, which requires close coordination between the application designer, the database implementer, and the operational architect in a team with different skillsets. Based on the CAP theorem, in a distributed cloud solution environment, when the partition tolerance condition must be met and the choice must be between consistency and availability, if the availability column databases are chosen, only weak consistency can be achieved. The rest are the strong consistency databases. See Table 6-12.

Table 6-12. *Distributed Database Selection Based on CAP*

Cloud Database	Consistency	Availability	Partition-Tolerance
Dynamo		X	X
RIAK		X	X
HBase	X		X
BigTable	X		X
Cassandra		X	X
Spanner	X		X

In general, key architectural considerations are:

- Driven by leading principles and business environments that impact enterprise success

- Determined by non-functional requirements that have significant service-level characteristics

- Decided individually or collectively by the seasoned, hands-on architect(s) or team member(s) who will implement the architecture in the decision area

- Validated by simulation or a subsequent review process

- Made transparent to key stakeholders and effectively communicated to them in plain terms prior to the decision

- Revisited when a solution architecture changes substantially

Business Decisions

Many architecture decisions are actually the result of business requirements and business decisions. Tradeoff analysis is often beyond the architectural scope when reconciling opposing forces between quality requirements and business constraints. For example, a business decision may replace an architecture decision:

- The identification and rating of significant services and business processes are left to business decisions.

- An architecture decision (such as a distributed database solution) may result in business resiliency, but architectural complexity and code size increase; it's up to the business to hire qualified professionals or outsource to third parties to implement the architecture decision.

- Many service-level decisions (including service degradation and volume restriction) are up to the business, as not all service requests can, should, or will be granted (e.g., in a highly distributed application environment).

- Performance and recovery targets are subject to business decisions.

A-ESA intuitively considers the business decisions based on the value of an overall strategic direction, capability model, business model (value stream model, operating model, and organization model), architecture model, governance model, and econometric model. Thus, the key considerations/choices take into account both architecture decisions and selective business decisions (likely from the Enterprise Decision Model).

Architecture Decision Validation

Big architecture decisions about technologies must involve technical specialists. Note that experience counts in the decision process.

The proof of the decision is in the validation. For example, given the complexity of distributed database selection (see Table 6-12), a mere product evaluation, no matter how detailed, is likely insufficient. Further consideration of the candidate database middleware in the solution context must be well thought out, followed by model validation and/or simulation testing.

From a simulation test (see Table 6-13), distribution architecture has improved throughput, yet the response time is longer (contrary to what some people expect). This explains that distributed transactions will improve scalability (or flexibility in a general term), but will not necessarily improve response time, which would be another issue as a tradeoff.

Table 6-13. *Simulation Test for the Distributed Database Consideration*

RQ-NFR Property	Synchronous Test	Async Test
Transaction rate	4903.88 trans/sec	6180.44 trans/sec
Response time	0.03 secs	0.09 secs
Throughput	27.18 MB/sec	38.40 MB/sec
Elapsed time	60.22 secs	60.10 secs

ESA project experience has proven that bigger spends, higher technology, and more skills don't naturally lead to better solutions. Architectural tradeoffs require architectural thinking to make balanced, larger, *sweet spot* decisions.

HINT

Deferred Decision: Not all architectural decisions need to be made now. They can be made for the future, or at the appropriate time. This avoids over-architecting or over-optimizing, focuses on the important decisions at the current stage, reduces technical debt due to uncertainty, improves efficiency and reduces costs, and makes decisions when sufficient information or feedback from key stakeholders is available.

Comparative Rating Techniques

Using a simple rating is an efficient comparison mechanism, sometimes more powerful than complex calculations. The rating covers a wide range of criteria.

Table 6-14 shows an illustrative NFR rating for an overall solution during a tradeoff analysis. When Importance means much more than Urgency, Importance prevails.

Table 6-14. *Simple Comparative Rating on NFR*

A-ESA Notation	Importance (I)	Urgency (U)	I + U	Efforts
[RQ-AGILITY]	4	3	2.3	4
[RQ-PERFORMANCE]	4	2	2	3
[RQ-SCALABILITY]	5	3	2.6	4
[RQ-SECURITY]	3	1	1.3	4
Note: − Rating: 1 (lowest) to 5 (highest). − A weighted average can be applied depending on the solution needs. − *Likelihood* is also a common factor to include. − For some solutions, a range of metrics (worst, best, median, etc.) needs to be specified for a more meaningful comparison.				

For the complex and critical tradeoffs between a set of requirements or between requirements and constraints, you can leverage various evaluation techniques for the initial analysis (see Table 6-15).

Table 6-15. *Techniques for Initial Tradeoff Evaluations*

Technique	Description	A-ESA
Cost-Benefit Analysis	Consider the cost and value of meeting each requirement.	[VL]
Pareto Principle	Apply the 80/20 rule to focus on the most important things that will have the biggest impact while managing constraints.	[PR]
Risk Assessment	Consider the risks of not meeting a requirement or exceeding a constraint.	[RK]
Constraint Identification	Identify all constraints, including technical, budget, regulatory, and resource limitations.	

(continued)

Table 6-15. (*continued*)

Technique	Description	A-ESA
Requirement Prioritization	Rank requirements based on their importance to stakeholders.	[RQ]
Case Scenario Analysis	Create different case scenarios to understand the impact of different decisions.	[UCM]
What-If Analysis	Simulate how different tradeoffs affect the system or product.	[VLD]
Prototyping	Test how different choices affect the system's performance and usability.	[DEV]
Utilization of Tradeoff Curves	Use plot tradeoff curves to visualize the relationship between requirements and constraints.	[MW]
Economic Model	Use models to predict how different choices affect the finances and goals.	[ML]
Multiple Criteria Analysis	Use formal analysis such as Multi-Criteria Decision Analysis (MCDA), Analytic Hierarchy Process (AHP), and Technique for Order Preference by Similarity to Ideal Solution (TOPSIS).	
Stakeholder Interview and Workshop	Speak with stakeholders to learn what they think is important and what they can accept.	[KC]
Matrix Analysis	Use a matrix to show how requirements and constraints affect each other.	
Root Cause Analysis	When there is a problem, find out why and look at ways to fix it.	
Balanced Scorecard	Consider multiple perspectives when making tradeoffs.	
Sensitivity Analysis	Determine how the system or product reacts to changes.	

Guide 4: Application Architecture Techniques

Enterprise-level application architecture is the most challenging part of the overall enterprise solution architecture, as it spans functional and operational aspects and is closely related to business and data architectures.

Application architecture not only matters to the traditional architectures, but also to the microservice architectures. Note that microservice architecture covers both the application level and the operational aspect. Application architecture is the foundation for microservice applications, and a holistic view is a must for a sound microservice architecture.

This section describes proven partitioning techniques for IT architectures, particularly application architecture. See Example Solution Profiles 5 and 6 for illustrated service-based application architecture and microservices domain-driven architecture.

Serviced-based Application Technique

A-ESA's IT service-based application partitioning techniques are distilled into five application design principles, a guiding principle, and a governance compliance (see Table 6-16), and they are well applied with corresponding A-ESA elements (see Table 6-17). These techniques plus appropriate patterns form well-constructed software applications.

Table 6-16. *A-ESA Guidance Techniques of Application Partition*

Technique	Statement	A-ESA Notation
Abstraction	Application architecture should hide the working details of subsystems or services.	Guiding Principle [PR]
Granularity	*Application granularity should strike a balance between coarseness and fineness.*	Principle [PR]
Grouping/Layering	*Application should be properly grouped or layered.*	
Cohesion	*Application services should be tightly cohesive.*	
Coupling	*Application service should be loosely coupled.*	
Autonomy/Isolation	*Application services should stand on their own.*	
Separation of Concerns (SoC)	Architecture should be unglued into parts, each of which may share the same service-level characteristics.	Governance [GV]

Note:

– *Italics* indicate five application design principles.

– The Agile application principles cover more than meets the eye. For example, layering includes domain, and autonomy includes technical isolation.

Table 6-17. *A-ESA Realization Techniques of Application Partition*

Technique	Realization Description	A-ESA Notation
Grouping/ Layering	In A-ESA, there is no layer element, which is covered by the domain, group, and view frame elements. Generally, domain means physical grouping while group means logical grouping. View-frame is a drilldown layer view.	– Domain [DM] – Group [GP] – View-frame [VF] – Pattern [PTN]
Granularity	In general, granularity is embodied by both coarse-grained services (domain or application services, etc.) and less-coarse-grained services (functional services or operational services, etc.).	– Domain [DM] – Application [AP] – Bounded Context [BC] – Service [GS] – Service Component [SC]
Cohesion	Cohesion is embodied by the responsibilities of service component, or functional services.	– Functional Services: [UI], [AS], [DS] – Service Component [SC]
Coupling	Coupling is embodied by the connection elements between services.	– Connections: [AN], [FW], [CN], [RN] – Views: [PFV], [PRM], [SIV], [DEP]
Autonomy/ Isolation	Isolation is enforced by dividing services into four architectural spaces (application logic, data, tech, and user interface) and a service component. Autonomy (self-contained and isolated) is applied to a deployment package unit.	– Functional Elements: [UI], [AS], [DS], [TS], [SC] – Service Interface [SI] – Deployment Package [DP]
Separation of Concerns (SoC)	Encapsulation of information within a service is embodied by a well-defined service interface.	– Service Interface [SI]
	Independence is embodied by the connection elements.	– Connections: [AN] [FW] [CN] [RN]
	Grouping of services with the same service-level characteristics is embodied by a deployment package.	– Deployment Package [DPM]

The application principles focus on application/functional characteristics (more from *software engineering*), but they strongly influence technical/operational characteristics (more from *systems engineering*) and design decisions. Note that A-ESA functional services follow the five principles at the architectural level. For service component design, you can optionally use the DDD (domain-driven design) principles for cloud-native microservice design and a software design principle for object-oriented design.

For ease of recall, all the principles can be loosely grouped into two CAGs. The first CAG (Cohesion, Abstraction, and Granularity) leans toward content, and the second CAG (Coupling, Autonomy, and Grouping) leans toward context. So the two CAGs can be further condensed into two Cs—Content and Context—or from a design perspective—Cohesion and Coupling.

HINT

SOLID Principle: This is a software design principle that includes: Single Responsibility (high cohesion), Open Closeness (isolation/coupling), Liskov Substitution (cohesion/coupling), Interface Segregation (cohesion/isolation), and Dependence Inversion (layering/coupling).

Abstraction

According to the Cambridge dictionary, *abstraction* is the quality of being very general and not based on real examples.

Abstraction in application architecture serves as a *guiding principle* for the rest of the application techniques. In particular, the level of model abstraction determines its application granularity.

A-ESA looks at the big picture. Experience shows that any model-driven architecture without a certain level of abstraction can never lead to a successful delivery of a (large and complex) enterprise solution or produce a useful architecture model.

HINT

Software Abstraction: From a software design perspective, abstraction is the number of abstract classes and interfaces divided by the number of classes in a component. It ranges from 0 to 1. The larger the division, the greater the abstraction.

Grouping and Layering

According to a well-known definition, a layer is a subsystem that provides services at a higher level of abstraction. In A-ESA, a layer can loosely mean a scope or a group, or a conceptual domain. Layering and grouping can often be used together to create a more organized and maintainable system.

Here are a few points about layering:

- **Closed layering** means that a given layer can only access the services in the next immediate layer, while **open layering** means that a given layer can access any of the layers.

- **Strict layering** means that a well-formed layer can only depend on lower layers. A less strict open layering allows dependencies on lower layers.

- **Tier** means a physical partition or grouping, or a physical layer.

- The **application layer** acts as an API, while the **domain layer** (business logic layer) acts as an implementation of the API.

- Layering is not conducive to an Agile and distributed environment. However, domain layers or event flow layers (producer, consumer, and channel/router) still need to be specified.

In short, layering follows the principle of *generality*, i.e. abstract services tend to stay in the lower layers. In general, there will always be some degree of layering, depending on the architectural style of the solution. Closed layering is generally required for good architecture. Layering helps prevent enterprise hairballs. In practice, layering is the *most powerful technique* because the level or style of layering dramatically affects other techniques.

HINT

Layering vs. Grouping: Layering is about creating a clear separation of concerns through a vertical stack of components, while grouping is about organizing components into logical units that are related by functionality or domain.

Sinkhole Anti-pattern: This describes the situation where requests flow through multiple layers as simple pass-through processing. If 80 percent of the requests are this type of simple pass-through processing, open layering is considered.

Granularity

The *toughest application architecture technique* is granularity because there is no best approach to getting the right level of granularity. The definition of "right" depends on your business flexibility and the architectural balance between complexity and adaptability.

The concept of granularity is relative and only becomes meaningful when comparing systems or modeling of systems. In A-ESA, granularity-wise, System Services > Application Services > Domain Services > Functional Services > Service Components. Granularity can be specified by bounded contexts (domain contexts implemented in microservices) and functional services. The A-ESA IT service-based modeling naturally incorporates the granularity into the architectural notations to facilitate good architectural work.

Granularity is a key attribute resulting from the way reality is abstracted in the modeling process. While maintaining *holism* in the architectural abstraction process, you can also apply *reductionism* to break a large service into smaller services until it fits into the application architecture. The determination of service granularity depends on several factors simultaneously, such as:

- Functional calls of each service

- The associated data update of each service

- The service definition from a given architectural style

- The number of service components exposed by a service interface

Here are a few terms that are different but closely related to granularity:

- Abstraction

  - Concerns: Granularity is concerned with how well information or systems are decomposed, while abstraction is concerned with simplifying models to understand and handle complexity.

  - Purpose: Granularity may be used to provide different levels of detailed views, while abstraction is used to create models that are easy to understand and manage.

  - Detailed level: Granularity may contain more detail, while abstraction reduces detail for easier processing.

  - Application: Granularity can be used at all levels, from macro to micro, while abstraction is often used for high-level architecture and design.

  - A-ESA context: Abstraction is a general concept with a wider broader scope, while granularity is more narrowly defined to service or component.

- Domain

  - Domain terms vary from team to team.

  - The granularity of a bounded context or domain entity may vary in size and by the number of attributes defined by different domain experts and business intents in a large-scale application architecture.

- Modularity

  - Modularity means dividing systems into parts that can change without forcing changes in other parts.

  - Coarse-grained systems have fewer and larger discrete services or components than fine-grained systems.

  - Unlike component modularity, service granularity is more subjective and based on domain knowledge.

HINT

Modularity from a Software Architecture Perspective: Modularity can be measured by the Modularity Maturity Index (MMI),[4] which determines the amount of technical debt in an existing system through a weighted calculation: pattern consistency (25%), domain and technical modularization (25%), technical and domain layering (15%), class and package cycles (15%), internal interfaces (10%), and proportions (of lines of code) (10%).

Cohesion

Cohesion is a state of holding together. If a service or service component fulfills the same type of responsibilities as deemed appropriate in the solution setting, it is cohered. Otherwise, for example, if a service fulfills both a payment responsibility and a delivery responsibility, it has bad cohesion.

---

[4] Vide: Sustainable Software Architecture/Lilienthal, Carola. dpunkt.verlag, 2019

In general, *high cohesion* is required for good architecture, but there will be exceptions when considering key NFRs. Note that A-ESA can focus on either service-level or service component cohesion.

HINT

God Object: A god object is an object that has access to a significant amount of data and is capable of performing a wide range of operations. Except on a high-performance platform, avoid god-object-like cohesion for service components.

Coupling

Coupling is the degree to which separate parts of the system are interconnected. *Loose coupling* is generally required for good architecture, but there will be exceptions. There is a strong correlation between coupling and other system quality attributes, such as complexity, adaptability, modifiability, and reusability.

Coupling is closely related to dependency and cohesion. In general, cohesion and coupling are mutually exclusive. Coupling can be measured by the average number of service dependencies, inbound or outbound. Note that cyclic dependencies are not allowed.

There are various degrees of coupling, including data coupling (via elementary data), stamp coupling (via composite data), control coupling (via internal control logic), global coupling (via static data), content coupling (via internal data change), and fusion coupling (via both data and control logic). A-ESA does not need to specify these detailed coupling mechanisms, but can optionally specify a coupling requirement as a principle or rule.

HINT

Law of Demeter: In software design, coupling follows the Law of Demeter (LoD) principle, or the "don't talk to strangers" rule, which encourages information hiding, abstraction, and narrow interfaces.

Autonomy and Isolation

Autonomy is measured by how each service stands alone and operates independently. It should also be technically isolated or physically segmented. Autonomy is generally required for good architecture, but there will be exceptions depending on the architectural style.

Service isolation can be achieved by separating the business logic from the technical implementation via independent deployment, interface façades, communication adapters or connection bridges, control mediators, and so on.

HINT

Autonomy vs. Isolation: Autonomy in IT is about independence and self-operation, while isolation is about separation and protection. Isolation and autonomy can coexist in certain scenarios. While designing a system with a high degree of autonomy often requires some degree of isolation, designing an application with isolation does not necessarily require autonomy.

Separation of Concerns

The above techniques are interrelated and should be applied in a balanced manner. Their purpose, in a sense, is the separation of concerns, which is more associated with loose coupling and high cohesion.

Separation of concerns means that each part of the system solves a problem. In A-ESA, the separation of concerns is enforced by the dependency rule, the service interface specification, and the deployment package mapping.

Example of Application Partitioning Practices

This section illustrates some application partitioning practices. As a form of partitioning, decomposition occurs in the business logic, as well as in the frontend, backend, and integration layers.

General Application Partitioning

Application partitioning is one of the mechanisms used to break a large system into smaller pieces. Table 6-18 illustrates the technique in an *order management* application.

Table 6-18. *Variations of Application Partition*

Technique	Variation	Situation	A-ESA Notation
Granularity	Coarse-grained	When a system has a small volume, it has highly coarse-grained functional services such as customer, order, and delivery in a monolithic architecture.	– Application Services [AP] – Process/Composite Services [PS]
	Fine-grained	When the system grows large, it has more granular functional services such as order and shopping cart.	– Functional Services [UI], [AS], [DS], [TS] – Microservice Services [MS]
	Mixed granularity	When the system needs to implement scalability and resiliency to meet burst or peak volume, it needs distributed granular services, including microservices, although a single microservice is not necessarily granular.	– Domain [DM] – Application Services [AP], including Microservice [MS] – Functional Services [UI], [AS], [DS], [TS]
Coupling	Business coupling	Such as order, shopping cart, payment, and fulfillment	– Bounded Context [BC]
	Functional coupling	Such as loyalty, coupon creation, fraud prevention, coupon usage, and coupon issuance	– Functional Services [UI], [AS], [DS], [TS]
	Deployed service coupling	Such as service dependency or communication framework	– Deployment Package [DP]

(*continued*)

Table 6-18. (*continued*)

Technique	Variation	Situation	A-ESA Notation
Isolation	Read/write Isolation	Such as – Writing data – Updating data	– Data Service [DS]
	Static/dynamic isolation	Such as – Static data (caching) – Dynamic (synchronization or replication)	– Data Service [DS]
	Data partition	Such as database partition	– Data Store [DB] – Data Schema [SM]
	Service interface segregation	Such as RMI, rest API, and web service	– Service Interface [SI]
Cohesion	SLC cohesion	Such as product details CDN, UI rendering, and CDN	– Middleware [MW]
	Service composition	Such as order logic process	– App Logic Service [AS]
Layering	Modular layer	Such as – UI view layer – Process control layer – Service bus layer – Data service layer	– Domain [DM] – Group [GP]
	Scaling layer	Such as – Vertical scalability – Horizontal scalability	– Domain [DM] – Pattern [PTN]

Note:
– Application partitioning is closely related to data architecture and can go hand in hand with database partitioning.

Software developers are familiar with the well-known SOLID principles. These principles are somewhat similar to the application architecture techniques, although they focus less on architectural granularity and abstraction.

MSA Application Partitioning

Table 6-19 shows the A-ESA application architecture mapping from typical microservice application design principles. Since microservice architecture requires a DevOps environment, technical services are considered at the same time. As you can see from the MSA application principles in the table, they align well with A-ESA.

Table 6-19. *Implication of MSA Principles*

A-ESA Area	A-ESA Technique	MSA Principles	Implication
Application	Cohesion	High cohesion, single concern or responsibility	– Self-containment
	Coupling	Loose coupling	– Decentralization
	Isolation	Abstract away from any dependencies, with interface segregation	– Service contract – Statelessness – Composability – Deployability
	Granularity	Business-driven	– Reusability
	Grouping	Domain-driven design	– Discrete boundaries
Technical	Autonomy		– Automation and resilience – Discoverability

Note:

– The technical area is taken care of in the overall architectural consideration or technical architecture.

Application Component Interface Specification

For A-ESA, signification service interfaces or service component interfaces must be specified. Table 6-20 shows an example technique for specifying service interfaces.

Table 6-20. *Service Interface Specification Techniques*

Technique	Example Description	A-ESA Notation
Interface specification from use case	For example, from the shopping cart use case, we can identify the following candidate interfaces: – Select shopping list – Create shopping list	– Use Case [UC]
Interface specification from interaction view	For example, in a fulfillment service interaction view, we can identify candidate interfaces from the verb actions: – Reopen order – Submit order	– Service Interaction [SIV]
Interface specification from business entities or data tables	Such as order, delivery, and payment account	– Data Service [DS] – Table [TB]
Interface specification from user interaction	Such as order user interface	– User Interface [UI]
Interface specification from existing application or systems	This is a bottom-up approach common for legacy system access or integration, such as ERP supply chain system interfaces	– Application [AP] – System [SY]
Interface specification from middleware	Such as an interface to order database	– Middleware [MW]
Interface specification from process	Either orchestration or choreography such as an order process	– Process [PS]

(continued)

Table 6-20. (*continued*)

Technique	Example Description	A-ESA Notation
Specification of information state	Identify information types that are in-flight or non-persistent, such as transaction type, amount, etc.	– Data Service [DS] – Domain-Transaction [TR] – Information Relationship [IRV]
Identification of pre- and post-conditions for service contract	Pre and post conditions: – Pre-condition for the payment, for example, the transaction must be created and the balance has been deducted – Post-condition: The guaranteed operation if the pre-condition is met	– Service Interface [SI]

Application Domain Structuring (Stages)

The application domain service structuring process typically follows three stages, as shown in Table 6-21. Because the structuring process is iterative and dynamic, the steps are not strictly sequential in reality. See Figure 2-33 and Table 2-14 in Chapter 2 for a modeling example of the structuring process.

Table 6-21. *Domain Service Structuring Stages*

Stage	Step Coverage	Description	A-ESA View
Identification	Layer partition	Determine service layers. You can skip this if the solution is for API services or the like	– Architecture Overview [AOV]
	Problem domain	Determine domain services	– Use Case Scenario [UCM] – Business Process [PRM]
Specification	Solution domain	Define service component domains. You can skip this if the solution is for a monolithic application	– Service Relationship [SRV] – Information [IRV]
	Solution domain service	Specify functional services	– Service Relationship [SRV]
Realization	Component structural relationship	Determine cohesion (similar responsibilities)	– Service Component Realization [SCR]
		Determine component dependencies (loose coupling)	– Service Component Realization [SCR]
		Detect common codebase	– Service Component Realization [SCR]

Note:

– The structuring phase approach applies, with slight variations, to traditional monolithic solutions and service-based architectures.

Microservice Domain Structuring (Steps)

Although there is no mechanical process that will produce the "right" microservice design without careful consideration of your business goals, business domain, and requirements through the modeling process techniques, Table 6-22 helps guide and accelerate the microservice architecture with a step process. See Profile 6 in Chapter 2 for a modeling example of the structuring steps.

Table 6-22. *Microservice Design Steps*

MSA Design Task	Description	A-ESA View
Analyze the domain	Use a DDD approach to map all of the business functions and their connections and points of integration with external systems.	Use Case Scenario View [UCM], Process View [PRM]
Identify bounded contexts	Group functionality according to whether different functions will share a single domain model.	
Define entities, aggregates, and services	Apply microservice patterns to identify natural boundaries for the services in an application.	Functional Service View [SRV], Interaction View [SIV]
Specify microservice boundaries	Specify microservices by analyzing bounded contexts, aggregates, and domain services.	
Consider key NFRs	Decide on NFRs based on microservice principles.	Metrics [MTS]
Choose data store and determine data collaboration	Carefully consider: – How to propagate updates across services. – How to manage eventual consistency.	Data Info Service [IRV], Pattern [PTN]
Deployment walkthrough	Walk through the deployment environment to identify hotspots.	Operational View [DEP]

Note:
– These are the common steps in a microservice design, which is an iterative process.

Here are some empirical architectural techniques when considering microservice application architecture and transition:

- A true microservice is a bounded context that, upon transformation, becomes an architectural quantum. It's an *independently deployable* component with its own private data source and highly cohesive functions. Stand-alone edge services (such as utility services), weak consistency services, or externally low-coupling services (such as an independent application service) are good candidates for microservices, which are not necessarily micro in size (lines of code). Examples of candidate application-level microservices include logins, authentications, payment processes, notifications, and search components.

- A microservices application has more *moving parts* than a monolithic application. A single service is probably simpler but the entire microservice solution is more complex. A service-based architecture (Solution Profile 5) is a *middle-of-the-road solution* because it reduces the quantum size while also reducing the coordination and transaction headaches, especially in a complex application process environment. To transform from monolith to microservices, start with a transformation to service-based architecture.

- To transform to a microservice architecture, start with services that have fewer dependencies and easy decomposability, follow the DDD principles, and apply patterns such as Strangler Fig and Anti-corruption Layer for *multi-phase transition* and business continuity. For the introduced data synchronization issues, use a shared database and apply patterns such as Database Wrapping Service and Database View.

Note that it is not always possible to fit a certain elements of a complex system into the predefined, structured parts of a domain model.

HINT

Economies of Scope vs. Scale: A microservice architecture is good for "economies of scope" when it is more cost effective to produce wider solutions or services in tandem, while a monolithic architecture is a better choice for "economies of scale." The cost effectiveness of an MSA is achieved through modularity, flexibility, reusability, specialization, technology diversity, rapid iteration, innovation, and risk management.

Architecture vs. Design: In a sense, solution architecture is design, but not all solution designs are solution architectures.

Application Data Service Structuring

Table 6-23 shows a critical data domain structuring process in an A-ESA model. See *Data Domain Service Determination* in Profile 5 (in Chapter 2) for a modeling example of the structuring process.

Table 6-23. *A-ESA Mapping to Data Domain Structuring Processing*

Step	Description	A-ESA Notation
1	Define the data domain, if any.	Domain [DM]
2	Decide on data integration or distribution.	Key Choice [KC]
3	Determine predefined or dynamic schemas.	Data Schema [DM-SM]

When structuring the application data service domain, keep the following in mind:

- The data source must always be accessed by a data service.

- The writing data service usually owns the data source.

- Direct access to the database from application logic services or technical services should be avoided.

- Exceptions (e.g., for performance reasons) should be made with extreme caution.

In complex scenarios, data domain structuring must consider the overall information and data architectures in order to provide guidance for data services and cross-cutting integrations, including:

- Provide data design guidance for entity-relationship modeling, data modeling, normalization, data classification, and metadata management.

- Ensure data consistency and distribution in master data management, data integration, cloud and distributed systems, event-driven architecture, and Data as a Service (DaaS).

- Ensure compliance with data security and relevant privacy regulations, such as HIPAA or CCPA,[5] and data governance.

- Ensure appropriate technical data services, data analytics, data lifecycle management, and so on.

HINT

Application vs. Technical Data Service: Application data services are solution-specific and require application logic design and implementation for the data services, often with framework support, while technical data services often use middleware or platform services with configurations to fit the data needs of the solution.

---

[5] The California Consumer Privacy Act

Guide 5: Technical Architecture Techniques

Unlike the application architecture, the technical architecture is relatively independent of the application logic. However, technical architecture can change the way the application or data is architected. Emerging trends in technical architecture impact techniques, such as cloud-native elasticity, DevOps continuous delivery, hybrid cloud, GitOps, Infrastructure as Code (IaC), infrastructure automation, and chaos engineering.

Despite these changes, A-ESA remains committed to the overarching concerns of business application usability and data reliability. These concerns transcend technologies and technical platforms. In A-ESA, deployment package mapping is a key link between application architecture and technical architecture when connecting the software development environment to the system runtime platform. A technical architecture can be built from experimentation, capacity planning, or non-functional requirements based on IT strategies. In addition, a technical architecture can be prioritized based on business impact and cost effectiveness.

Deployment Package Mapping

Although the technical architecture stands relatively alone, the deployment package mapping connects the technical architecture to the application and data architectures. *Deployment mapping* means that functional services and microservices must be deployed based on service-level characteristics (such as access frequency, concurrency, volatility, usage intensity, bandwidth state, composition, technical isolation, etc.). A well-constructed service often does not mean an SLC-compliant deployment package except in a microservice architecture, when it means both a service and a deployment package/container. Table 6-24 shows illustrative mappings.

Table 6-24. *Deployment Package Mapping*

Functional Service	Deployment Package	Mapping Relationship
Multiple functional services	One deployment package	One to many (e.g., SLA matching)
One functional service	One deployment package	One to one (e.g., microservice)
One functional service	Multiple deployment packages	Many to one (e.g., replication)

Note:
- Deployment package mapping is not a one-way street. It will have an inverse impact on the application architecture.
- Depending on the architectural style, the deployment package can refer to a container environment.

NFR-driven Operational Excellence

The technical architecture, or operational architecture, can be driven by non-functional requirements. The technical architecture is easily determined by requirements, proven cases, and leading practices. A-ESA is more concerned with the operational side of technical architecture. Operational architecture is more about how the system supports availability, monitoring, backup, and so on. Table 6-25 provides an example of techniques based on key NFR attributes, including the runtime availability and non-runtime manageability of an NFR-driven operational architecture.

Table 6-25. *Techniques Based on Key Attributes*

NFR Cluster	NFR Metric	Availability Technique	
		if 99%	if 99.99%
Availability	Scalability	Vertical scaling	– Auto scaling for application and data store – Read-data replication
	Elasticity	If failure, build up from scratch	– Auto-scale – Web and application self-healing – Fault-tolerance
Manageability	Deployability	Runbook for deployment and rollback	– Automated canary/blue-green deployment – Automated alert-based rollback
	Observability	Health checks	– Health checks – Alerts on specified metrics and failures
	Recoverability	Runbook for backup and restore	– Automated backup – Automated recovery based on the systematic incident simulations

A cloud technical platform architecture, as a self-managing architecture, typically provides menu-like choices for different NFR needs. It is more flexible to scale in, scale down, scale up, or scale out from a single region to multiple regions. Even with cloud and reusable technical architectures, there still needs to be a clear value proposition and some principles, as well as applicable case scenarios. For example, while the cloud technical architecture can save a lot of headaches, there are tradeoffs such as cost and ownership. Therefore, the full set of A-ESA principles still applies to different technical architectures, including cloud hosting.

Technical Data Service Structuring

Data services in a distributed architecture are a cross-cutting concern that goes beyond business and application architecture. Data services are also closely tied to the technical architecture. For technical data services, you can typically choose a distributed database solution and/or a highly reliable platform solution for stateful data needs.

There are many distributed transactional databases, including Google's Spanner, PingCAP's TiDB, Tecent's TDSQL, Alibaba's OceanBase, CockroachDB, and VoltDB. See Table 5-25 in Chapter 5 for the technical architecture style of distributed transactions. Using a distributed database is a big architectural decision. Table 6-12 shows the considerations for selecting a distributed database. The selection decision is based primarily on consistency, availability, performance, security, and maintainability. A distributed database requires more effort and a complex environment with platform considerations. It goes beyond a generic middleware definition and requires a clear operational view for proper deployment and integration.

In addition to distributed database solutions, you need technical data architecture techniques for data state considerations based on your chosen architectural style, solution scale, and business objectives to achieve cost effectiveness. Effectiveness can only be achieved through a multi-dimensional consideration of all impacting system elements. Table 6-26 shows the key architectural techniques used for a well-known case of highly reliable stateful website systems across clients, APIs, and servers.

Table 6-26. *Techniques for Highly Reliable Stateful Systems*

A-ESA Notation	SLC Policy	SLC Behavior	Governance Technique
Reliable Stateful Clients			
Service Level Objective [RQ-SLO]	Service level objectives (SLOs) per namespace and type	N/A	Base SLOs on the particular namespace, such as namespace and endpoint pair ([NS1, GET], [NS1, PUT], [NS2, SCAN]).
	Concurrency limit hedging and GC-tolerant timeouts	N/A	Use concurrency limiting to prevent too much load going to the backend services.
	Retries and load balancing	N/A	Retry (relevant to the SLO target) and use the load balancing algorithm: weighted-choice-of-n.
Client [FE]	Servers signal to clients	Service time out	With service level objectives, use server signaling to convey context to clients to improve reliability.
Reliable Stateful APIs			
Service Interface [SI]	Rule for idempotency tokens	Level of consistency by idempotency tokens	Depending on how strongly consistent, use client monotonic timestamps or take a global transaction ID.
API [SI]	Key value abstraction	Streaming API issue	Turn stateful APIs into paginated APIs that return fixed amounts of data.
Event Interface [SI]	Time series abstraction	Consistency tradeoff	Based on the different use cases, offer three modes to clients: fire and forget, enqueued, or durable.

(*continued*)

Table 6-26. (*continued*)

A-ESA Notation	SLC Policy	SLC Behavior	Governance Technique
Reliable Stateful Server-end Operations			
Data Service [DS]	Same availability and different reliability by SLOs (failure frequency and recover speed) to reduce blast radius	SLO-A service always fails a little bit; it never recovers.	Require request hedging or retries.
		SLO-B service occasionally fails cataclysmically. It recovers quickly.	Need load shedding or backpressure.
		SLO-C service rarely fails, but when it does fail, it fails for a long time.	Need faster detection and failover.
	Spend on stateful, save on stateless	Stateless services can't autoscale quickly	Reserve 33% headroom for failover (buffer for slow scaling services and pre-inject it for fast scaling services).
	Stateful image agility	Problematic stateful service that is coupled to the running hardware	Carve mutation seams into the stateful images to react more rapidly without affecting the system's availability.

(*continued*)

409

Table 6-26. *(continued)*

A-ESA Notation	SLC Policy	SLC Behavior	Governance Technique
Caching [CA]	Different level of caching	Eventual consistency	Use near caches that live on service hosts.
	Cache in front of services	Fail fast of video metadata service using warming caches	Whenever the underlying data of that service changes, the service recalculates the cache value and fills that cache with the new view.
		Best-effort consistency	Use remote caches based on memcached.
		In-region read-your-write eventual across consistency	Use stateful databases with Apache Cassandra running in four-region full-active mode.
Server Node (ND)	Tenancy partition	Multi-tenant data stores that served multiple customers, with rare failures that had a significant blast impact	Use single tenants that don't share caches or data stores, resulting in more frequent failures but a more isolated blast radius.
System [SY]	Prioritized capacity cluster	Busting events	Use automated workload capacity modes to make the least-regret choice by spending on critical (Tier 0) clusters while saving money on less critical clusters.
Cluster [CT]	High replication for reliability	Consistency across four regions (12 availability zones)	Replicate with highly reliable writes and reads (using quorums to accept writes in any region and having three copies in each region).

(continued)

Table 6-26. (*continued*)

A-ESA Notation	SLC Policy	SLC Behavior	Governance Technique
Middleware [MW]	Proactive monitoring	Errors and latency for hardware and software	Monitor continuously to detect and recover from failures before they propagate to the high-impact stateful service.
DevOps [DEV]	Smart deployment	N/A	Review the simulation of system failure on an Apache Cassandra fleet and use the in-place imaging technique.

Note:

— The A-ESA elements listed are a high-level abstraction representing a group of elements with the same SLCs. The requirement element [RQ-SLO] can be a specific functional or operational element when a clear SLC is exhibited.

— This table of A-ESA mapping of summarized architectural techniques is based on the Netflix information presented at QCon San Francisco, October 2023. See the following links for details:

— https://www.infoq.com/presentations/netflix-stateful-cache/

— https://www.infoq.com/articles/netflix-highly-reliable-stateful-systems/

— https://netflixtechblog.medium.com/

As you can see, for large-scale technical data architecture, you need to take an abstracted view of the system, focus on the overlapping areas of the data and the technical platform, and weigh metrics to come up with the right set of governance measures. Preferably, a model view corresponding to the tabular view (see Table 6-26) would better reflect the relationships between these measures.

HINT

NFRs vs. Governance: NFRs must be mapped to governance measures. The NFRs for the highly reliable stateful system in Table 6-26 include workload efficiency and optimization, redundancy and heavy caching, data recovery, constant monitoring for changes, and reliable APIs for data consistency based on the concepts of idempotency and fixed-size units of work. These NFRs can be mapped to governance techniques through architecture views or matrices.

411

Guide 6: Effective Governance

Governance is a huge topic. It starts with corporate governance, enterprise architecture governance, business architecture governance, and then includes enterprise solution architecture or domain architecture governance, followed by design and implementation governance. Governance includes cross-cutting topics such as compliance, security, and risk. For an otherwise broader discussion, the enterprise solution governance needs to be well aligned with the strategic model and enterprise solution management context (scope, resources, and finance).

A-ESA governance is characterized by architectural features such as governance metrics, guardrails (governance in action), or more specific governance functions. Governance standards such as SGVM[6] can be referenced, but a big governance approach should be avoided. For quality assurance, *architectural governance through modeling* is an effective and efficient mechanism for guiding and validating the architecture. A-ESA advocates an incremental process[7] for governance discovery and enforcement.

This section discusses enterprise solution architecture governance from the practical perspective of A-ESA.

Governance Authority

Depending on the nature and scope of the enterprise solution, the governance role may be a governance body, an architecture committee, a board of architects, an architectural authority (chief architect, lead enterprise solution architect, etc.), or Agile teams.

---

[6] The Open Group SOA Governance Vitality Method

[7] See examples of incremental governance in Chapter 5 of the book *Agile Enterprise Solution Architecture*.

In a *sprint* approach, the cycle is so short that it does not allow for sufficient governance. Instead, teams collaborate through accountability of significant measurements. Different levels of delegation can be used to define how governance is delegated between Agile teams.

In reality, there are many situations where each developer wants to make decisions in isolation, which is a major source of architectural and technical debt. Technical debt exists not only in the codebase, but it also pollutes the architecture; something that won't be seen in the short term. So, a *hero culture* and a *who-cares culture* are no-nos in a complex solution environment. The A-ESA model is a place for everyone to collaborate and cross-check. In A-ESA, the *owner role* is the *governance authority*. The authority governs things like how decisions can be rolled back or delegated to individual teams. If necessary, decisions can be made by a higher level architecture authority or by a larger group of enterprise Agile teams. Because modeling makes you think smarter, it can act as a supporting authority, especially in a junior team environment.

In any case, don't blindly follow an authority without a modeling approach. Key governance elements must be associated with a governance role for an A-ESA model.

Governance Enforcement

Governance is about enforcement. In A-ESA, governance enforcement is divided into functional governance and operational governance. It works with solution validation and governance management for architectural assurance and conformance. Even in a cloud hosting environment, development teams still need to follow functional governance and configure operational governance on the cloud platform.

The governance enforcement specification includes governance rule, function, and measurement. The governance rule focuses more on the *why* than the *how*. The calibration of the governance metrics can be fine-tuned to fit a particular solution environment.

Functional Governance

Governance function is one of the functional governance and includes fitness function,[8] pseudocode, statement, description, specification, or architecture-centric requirements that govern architectural characteristics.

Governance function is a powerful tool for architectural conformance, but granular specifications and overuse should be avoided. It is best to focus only on the architectural designs that are significant and exemplary, and leave the rest to the design and implementation teams.

Table 6-27 shows examples of governance function statements associated with A-ESA notations. See also Profile 5 in Chapter 2 for some governance function and fitness function examples.

Table 6-27. *Governance Function Statement*

Category	Governance Statement	Check	A-ESA Notation
Coupling	Check the number of relationships between functional services to avoid cyclic dependencies.	Auto	Service Relationship [SRV]
	There should be no more than three total dependencies for any service component.	Auto	Component Relationship View [SCR]
Layering	Define the distributed transaction boundaries to contain compensation errors within 1% of total transactions.	Manual	Pattern [PTN], Validation View [VLD]

(continued)

---

[8] As defined in *Building Evolutionary Architectures*: The fitness function provides an objective integrity assessment of some architectural characteristics that must be preserved.

Table 6-27. (*continued*)

Category	Governance Statement	Check	A-ESA Notation
Cohesion	Validate the overall level of indirection and nesting within a component to avoid cyclomatic complexity.	Auto	Service Component [SCR]
Modularity	Fitness function: <No component should exceed 10% of the total codebase>.	Auto	Service Component [SCR]
Note: – Those that can be automated are implemented using home-grown or commercial tools. – In A-ESA, layering includes the transaction domain.			

HINT

Governance Function Metrics: From a software architectural design perspective, there are many useful metrics that can be leveraged for governance functions when specifying *service components*, such as average component dependency (ACD), cyclicity and relative cyclicity, structural debt index, maintainability level, indentation debt, change history metrics, change frequency, code churn, and so on.

Operational Governance

On the operational side, the governance function also has a role to play. Similar to functional governance, operational governance has a number of specification mechanisms. It is more likely to take the form of a rule, configuration, or virtual service.

Architecture Validation

Architecture validation is part of the governance process. The obvious way to finally prove an IT architecture is to run it in a production environment, or to build and test it in a QA or DevOps environment. However, delaying the assessment and validation of an architecture until design and implementation defeats the purpose. The sustained delay is often at the root of many solution failures.

In fact, a large part of architecture validation is *model validation*, which checks for consistency, correlation, redundancy, complexity, transaction integrity, and interdependencies between governance measures and metrics. In a DevOps environment, architectural governance enables continuous modeling improvement based on feedback from testing and monitoring.

Importantly, architecture validation is not limited to testing and model validation. Proper use of *architectural techniques* and *tradeoffs* is an effective measurement approach. An expert opinion, a checklist, a scenario walkthrough, a leading practice, and a governance clue can all be effective validation.

HINT

Review vs. Validation: Review is closely related to validation. They are complementary. Review involves more discussion and provides feedback to improve solution quality and reduce problems during validation. In a traditional approach, validation is typically done at a later stage of the solution. In an Agile environment, validation is more frequent. Review and testing are more of a solution management concern.

Governance Automation

Much of the governance work can be automated, for example, modularity is easily automated as long as architectural governance rules are defined. Governance has capabilities for specifying automation work, extending playbooks[9] and runbooks[10] and managing the work that can't be computerized.

---

[9] Playbooks record the predefined steps to identify a failure scenario.

[10] Runbooks are a set of standardized written procedures for completing repetitive information technology (IT) processes within a company.

There are support tools for architecture governance, such as JDepend and ArchUnit, which are generally good tools for assisting architecture modeling. There are also many development tools for coding analysis and validation, such as SonarQube. For operational governance support, there are tools such as Jira Service Desk and SolarWinds from an ITAM[11] perspective. However, there are limitations in tool support, such as customization, architecture model integration, and design or technical focus only. Even an *AIOps* is limited in its input data and intuition.

Overall, the role of governance automation is still limited in many cases, and some manual work can't be eliminated given the nature of IT architecture where the *art* part plays a decisive role. Limitations include:

- Automation will never be a catch-all for architectural issues.

- Automation can't easily replace business expertise and architectural intuition.

- Architectural level quality, such as abstraction and granularity, can't be easily automated.

- Empirical experience, leading practices, high-fidelity simulation, and the like can precede automation for less costly and more effective results.

- Monitoring provides the runtime data for the solution, but it's still up to the enterprise solution architects to govern and improve the architecture model.

HINT

Governance Automation vs. Validation: Governance automation improves efficiency and effectiveness, while validation ensures that the architectural solution meets requirements and standards.

---

[11] IT Asset Management

Guide 7: Solution Management Context

The A-ESA modeling process is associated with solution architecture planning, solution project management, solution process, solution design, solution implementation, and solution maintenance. While this book focuses on enterprise solution architecture and its related modeling, *solution process* and *solution management* are also part of solution architecture governance.

Among people, technology, and process, people (including communication) are the most critical factor in solution management. A-ESA instills people concerns into the role element and organization view. The ESA architect works with the solution manager to reach agreement on *stakeholder management* by analyzing the organizational impact of human factors and clarifying RACI.

It must be made clear that A-ESA focuses on the *content* of the solution deliverable, while project management is about the project *context*, which focuses on the triangles involved: scope, schedule, and budget. In a sense, solution project management (which could be a whole topic in itself) is also a constraint for ESA.

In an Agile environment, solution architecture and project management are fused as a set of profile model views. Table 6-28 shows the solution project management definition and lifecycle management that can be captured and tracked as an integral part of an A-ESA model.

Table 6-28. *Solution Architecture Project Management in A-ESA*

Category	Task/Stage	A-ESA Notation	Related Property
Project definition	Project Objective	– Goal [RQ-GL] – Value [VL]	
	Key Issue and Challenge	– Risk [RK] – Decision Log [KC]	– Importance – Urgency – Effort – Feasibility – Cost – MoSCo[12]
	Stakeholder	– Organization View [ORG] – Role [RO]	– Executive Support – Project Manager if any – Architect – Team – Partner
	Scheduled Deliverable	– Deliverables [DL]	– External Dependency – Release Plan – Solution Milestone – Solution Timeline
	Acceptance Criteria	– SLA Requirement [RQ]	

(continued)

---

[12] Must, Should, Could

Table 6-28. (*continued*)

Category	Task/Stage	A-ESA Notation	Related Property
Project lifecycle mgt	Project Definition Attachment	− Architecture Overview [AOV] − Document File [DF]	− SoW[13] − Budget Estimation − Skill Assessment − Feasibility Study − Test Plan
	Project Start-up Environment	− DevOps [DEV]	
	Solution Environment	− Architecture Overview [AOV]	
	Solution Model	− A-ESA Model [ML]	− Model Notation Customization
	Project Planning for Scoping and Activity	− System Context [AOV] − View Frame [VF] − Task/Activity [TK]	− WBS[14] − Stage and Activities
	Project Monitor and Control	− Validation [VLD] − DevOps [DEV]	
	Project Closing	− Deliverables [DL]	

Note:

− As mentioned, A-ESA modeling can also be helpful to the solution breakdown structure.

− Deliverables include something intangible, like a customer satisfaction result.

---

[13] Statement of Work

[14] Work Breakdown Structure

In short, here are the points about solution project management in ESA:

- A-ESA takes a light view of project management, or operational and maintenance management.

- A well-formed A-ESA model contains brief, critical information about the role of people, technologies, and processes related to solution.

- Project management consideration is a natural modeling outcome, as A-ESA will balance between the cost/effort/skillset and benefits in the decision process.

- Agile ESA promotes a self-directed and blameless culture.

Architectural Assessment

Architectural assessment is an optional process when justified. An initial sketch modeling assessment might be an option. The assessment is a good approach to raise architectural issues (see Table 6-29) that might otherwise be known but are missed, unknown, and unrecognized.

Table 6-29. *Assessment-induced Sample Questions*

Assessment Area	Architectural Question	A-ESA Notation
Microservice Architecture	Does the service follow DDD principles?	Domain [DM]
	How will the distributed tracing be implemented?	Operational View [DEP]
Cloud Architecture	Does the platform support auto-scaling?	Cloud [CL]
	How does the platform handle the deployment units (virtual machines, containers, or serverless)?	Deployment Unit [DP]

For example, you can adapt different methods, frameworks, and approaches for different levels of architectural assessment:

- Architecture Tradeoff Analysis Method (ATAM)

- Software Architecture Analysis Method (SAAM)

- Aspectual Software Architecture Analysis Method (ASAAM)

- Cost Benefit Analysis Method (CBAM)

- TOGAF Enterprise Architecture Framework

- Microservices readiness or cloud migration assessment from your intended vendors, such as

  - Google Cloud Architecture Framework

  - Azure Well Architected Framework and microservices assessment and readiness

  - AWS Well Architected Framework

  - Alibaba Cloud Well Architected Framework

Architecture assessment often involves estimation, which requires different levels of effort: rule of thumb, tabular calculation, tools, simulation, and prototyping. Modeling is also a powerful estimation mechanism that requires relatively less effort. To achieve better estimation reliability, you can use model walkthrough or simulation. For example, for performance estimation, you need to be clear about capacity planning (application, hardware, network, resources, etc.), transaction volume, user activity (peak, burst, etc.), technical clustering (deployment mapping), and response time requirements (concurrency modeling). Keep in mind that a modeling estimate is only as good as the input data and the expertise. You can also resort to professional companies or organizations to perform the estimation and use the estimated results for validation or further modeling.

Requirements Management

Solution management includes requirements gathering and analysis.
It covers functional requirements and non-functional requirements.
For A-ESA, the focus is not on requirements gathering, but rather on
requirements mapping, which focuses on non-functional requirements.

There are several approaches to requirements screening, including
requirement chain and matrix (see Table 6-32 for techniques). A good
approach is to screen out significant requirements and map them to the
metrics elements in A-ESA. Table 6-30 identifies some common issues with
requirements mapping and lists the modeling resolutions.

Table 6-30. *Common Unaware-of Requirement Mappings*

Requirement Issue	Description	A-ESA Resolution
Narrow-minded view	Most of the non-functional requirements don't stand alone, and they need to be considered as a whole.	Metrics View [MTS]
	Use the boilerplate NFR template with caution, as each industry scenario may be dramatically different.	
Too many requirements	A set of interrelated requirements could be abstracted into a principle for effective enforcement and modeling.	Principle [PR]
	Not all the key requirements need to be written out, but they need to be linked in the model and properly governed.	Governance [GV]
Improper categorization	A functional requirement may also be a non-functional requirement (such as security or caching requirements).	Requirement [RQ]

Here are some rules of thumb about requirements mapping:

- Key requirements are not there all at once; they are built up incrementally during the modeling process. Look for hidden or implied requirements. NFRs can come from many sources: regulations/compliance, standards, industry norms, environmental variables, and so on.

- Any significant or prioritized requirements must be SMART.[15]

- Don't just rely on automated test scripts. Create test cases as early as possible for better requirements management.

Critically, identify requirements that are considered *significant* with the following traits:

- Critical business functions or processes that must provide continuous service or be complete with no exceptions

- Requirements mandated by key sponsors or influential users

- Requirements with high standards of security compliance

- Requirements that are part of integrated system(s)

- Requirements that affect a large group of users

- Business functions or processes with high financial value

- Requirements that are difficult or complex, such as high-impact transactions, scalability

- Representative solution case scenarios

---

[15] SMART commonly means specific, measurable, attainable, realizable, and traceable.

- First-of-a-kind innovations

- Requirements that will result in risk, budget overruns, or negative outcomes

- Urgent and high-impact business rules, constraints, and so on

Quality Measurement Indicators

Quality measurements are viewed from a different perspective in A-ESA. Table 6-31 explains the different measurement indicators and their connotations. As part of solution management, the number of metric indicators should be limited to avoid complexity.

Table 6-31. *Quality Metric Indicators*

Metric Indicator	Description	Audience Role
SLA (Service Level Agreement)	SLAs are generally derived from the Statement of Work or from the cloud service vendors. They are the agreed-upon requirements that must be met.	Solution provider and solution consumer
SLR (Service Level Requirement)	SLRs are intended or expected requirements.	Fundamentally, it is from the customers and users
SLO (Service Level Objective)	SLO is a predefined, specific, and quantifiable target that describes the performance or availability criteria of a service.	DevOps teams

(*continued*)

Table 6-31. (*continued*)

Metric Indicator	Description	Audience Role
SLI (Service Level Indicator)	SLI is part of SLM, and it usually refers to a SLO-specific metric, such as response time, success rate, etc.	Solution developer and operator
SLM (Service Level Metrics)	SLM is a metric used to monitor and measure the performance of a service and it tells whether a service is meeting its SLOs.	SA architect, solution developer, operator, QA, solution manager, IT governance team, and analyst
SLC (Service Level Characteristics)	SLCs or *architectural characteristics* are the key architectural concerns. They describe the performance characteristics of services.	ESA architect

Note:
- SLRs and SLOs are generally not legally binding, unlike SLAs.
- SLO is commonly used in a DevOps environment to describe the service metrics between teams. It can be considered a self-governance measure.
- Metrics can be expressed in different terms and derived from different approaches. For example, using GQM (goal-questions-metrics) to concretize a goal measure.

A-ESA is more closely related to SLAs, SLOs, and SLCs. In comparison, A-ESA is more concerned with SLCs (Service Level Characteristics). In practice, *measurement* SLCs are often considered within a specific range, such as peak, burst, critical case, worst case, best case, and so on. SLCs are used to define SLOs and are often part of SLAs. On the other hand, SLCs are typically the basis for solution architecture governance measures.

It is important to note that in A-ESA, SLC means more service level *characteristics or characterization* at the architectural level. It's a coarse-grained indicator of how a service, domain, or system performs, including how it behaves and reacts. In addition to functional or operational services, SLC is also taken into account by architectural services, frameworks, and platforms.

HINT

Metric vs. Measurement: These terms are used loosely in this book. Generally speaking, metric is conceptual and defines something to be measured for evaluation and comparison, and may be used for continuous monitoring and evaluation, while measurement is the practical application of metrics, generally to obtain data for a specific instance or result and may be one-time or periodic.

Techniques for Measuring Requirements

Measuring enterprise solution metrics, especially subjective qualities such as usability or adaptability, can present challenges. While ESA is responsible for requirement mapping, an initial and large set of requirements understanding and screening is part of solution requirements management. Various techniques or workarounds can be applied to overcome the limitations of subjectivity, context dependency, and variability (see Table 6-32).

Table 6-32. *Techniques to Work Around Measurement Limitations*

Category	Technique	Description	A-ESA
Survey and evaluation	User surveys	Survey users about their experiences and subjective qualities and use questionnaires that capture user experiences.	[TK]
	User personas	Create user personas and tailor metrics to reflect their needs.	
	Heuristic evaluation	Use heuristics or best practices to evaluate usability and adaptability.	
Metric dynamics	Longitudinal studies	Study how users' opinions change over time.	
	Multi-dimensional metrics	Create a set of usability and adaptability metrics and do not rely on one measure.	
	Technology watch	Keep up with new tech and trends to make sure your metrics are up to date.	
Testing	Usability testing	Test the usability with a group of users to gather data on how they interact with it.	
	A/B testing	A/B test different versions of an IT solution to see which is better.	
Analysis	Analytics	Use analytics to track user interactions and usage patterns.	
	Expert reviews	Use usability experts to assess an IT solution's subjective qualities.	
Design	Inclusive design	Include a wide range of users in the design process to account for cultural and individual differences.	
	Iterative design	Use an iterative design process that allows for continuous improvement.	

Another common issue in requirements management is the prioritization of requirements. Table 6-33 shows techniques or guidelines for prioritizing IT solution metrics in a resource-constrained environment.

Table 6-33. *Prioritization of Solution Metrics*

Guideline	Description	A-ESA
Define business goals	Understand the main goals and how IT helps achieve them.	[TK]
Develop an action plan	Make a plan for choosing projects and using resources. This plan should support the enterprise's main goals.	
Consider existing resources	Make sure the project fits your team's abilities and resources.	
Use decision analysis	In limited-resource situations, use Decision Analysis and Resolution (DAR) to compare classifiers and choose the best one.	
Consider performance metrics	Choose metrics that reflect how well your system works, how quickly it does its work, and how accurate it is.	
Select weighting priority criteria	Choose metrics that show how IT helps the business, such as cost, risk, quality, and revenue.	
Apply criteria	Use the priority matrix to apply the criteria, assign values, and score based on the sum.	
Analyze the results	Be aware that higher scores mean more important projects and make sure the results are reasonable.	
Conduct periodic reassessment	Update the priority list regularly to make sure the project is in line with the enterprise's goals and resources.	

A-ESA is primarily for key NFRs considerations, not for detailed requirements analysis or management. There are many tools and platforms that can help manage quality metrics more efficiently, such as Atlassian and Jira Service Management, DataFocusBI, and Yunxiao.

Cost Control and Management

Almost all NFR tradeoffs in ESA focus on economic costs, either in the short or long term. Economic costs include accounting costs (explicit costs) and opportunity costs (implicit costs). Broadly speaking, cost savings in ESA are achieved through one of two types of economies: economies of scale or economies of scope. An economy of scale means that costs are reduced by spreading fixed costs over more units of a single solution, while an economy of scope means that the average total cost of the solution decreases as a result of increasing the number of solution parts. The cost decision determines, for example, the architectural style or whether to move to the cloud.

Cost control and management extends beyond the solution architecture. It's the responsibility of the solution manager and the enterprise leaders to determine cost spend and budget allocation.

Cost is a major consideration in architectural tradeoff analysis. It is indirectly reflected in A-ESA views, elements, or properties (see Table 6-34). Most of architecture views and elements can have associated cost property attributes. An appropriate architecture will naturally reflect this.

Table 6-34. *Cost Constraint Consideration from Solution MGT*

Category	Objective	A-ESA Notation
Business	Align costs with business values and FinOps and establish a pricing model that includes cost baseline and service levels.	Value [VL]
	Align solutions with enterprise IT strategy and organizational efficiency.	Capability [CP], Value Stream [VL]
	Gather requirements and estimate costs.	Requirement [RQ]
	Consider costs from a business or customer perspective.	Case Scenario [PRM]
IT	Clarify accountability.	Role [RO]
	Identify risks.	Risk [RK]
	Ensure deliverables are within cost, schedule, and scope.	Deliverable [DL], Properties

The following sections discuss cost considerations for two inherent architectural requirements: sustainability and deployment economy.

Cost Impact on Sustainability

ESA's goal is sustainability, but cost is the primary constraint. Balancing the two is part of solution management. Table 6-35 shows the mutually influencing factors for such a balance consideration.

Table 6-35. Cost vs. Sustainability

Factor	Description	A-ESA
Cost-benefit	To cut costs, companies may look at whether sustainability projects are good for the environment or society while also saving money.	[VL]
Long-term value	Sustainability metrics focus on long-term value, while cost optimization helps organizations be cost-effective without losing sight of the long term.	
Supply chain management	Integrating sustainability metrics in the supply chain can save money by making logistics, resource management, and waste reduction more efficient.	
Return on investment	A sustainable project may have long-term economic benefits, even if the initial investment is high.	
Resource efficiency	Improving resource efficiency reduces costs while reducing environmental impact, which is an important part of sustainability metrics.	[RQ]
Innovation-driven	Sustainability metrics can encourage organizations to adopt new technologies and approaches, which may cost more initially but save money in the long run.	[CP]
Compliance costs	Compliance with environmental and social standards can be costly, but it is essential for sustainability.	[GV]
Risk management	By considering sustainability metrics, organizations can manage environmental and socially related risks, which may avoid future costs.	[RK]
Stakeholder expectation	Meeting stakeholder demands for sustainability can improve an organization's reputation and brand value, potentially leading to cost savings.	[RO]

Cost Impact on Deployment Approach

Another cost consideration for ESA is the deployment method. Deployment methods in ESA vary depending on development strategy, team size, skills, project complexity, and business needs. Making tradeoffs between cost and deployment method can also be part of solution management. Table 6-36 shows the costs and benefits of some common deployment methods. It is beneficial to consider the overall A-ESA considerations and solution management directions from management teams when making a decision about such tradeoffs.

Table 6-36. *Cost vs. Benefit of Deployment Methods*

Deployment	Cost	Benefit	A-ESA
A/B testing	Infrastructure needed to manage versions and analyze user engagement.	Decisions based on feedback and data.	[KC]
Blue-green deployment	Costs twice as much to maintain two identical production environments.	Deployments with no downtime and less risk.	
GitOps	Need to know Git and infrastructure as code and may need to train your staff.	Improved consistency, version control, and rollback.	
Waterfall deployment	Sequential phases lower initial costs, but may increase later costs if changes are needed.	Clear project structure and budgeting.	

Note:

– There is a long list of deployment methods, including DevOps CI/CD, serverless deployment, microservices deployment, Infrastructure as Code (IaC), hybrid deployment, dark launch, containerization and orchestration, canary releases, feature toggles/flags, rolling deployment, and short-cycle deployment.

HINT

ESA's Tasks in Cost MGT: There are two tasks that typically fall on the shoulders of the ESA architect in the solution environment: cost estimation and cost optimization:

- **Cost estimation** is part of the solution consideration, along with scope and schedule preparation. Although it's the responsibility of solution management, it often relies on the ESA architect to figure it out.
- **Cost optimization** is a major factor in the architecture decision. Since an ESA architect has less control over the budget, they will only architect within the budget constraint.

Test Management

Testing is not an afterthought, nor is review and verification. They are naturally part of the A-ESA governance and solution management process. The architecture-related issues discovered late in the solution lifecycle can be difficult to fix and require costly application rework. Even if you embrace *DevTestOps*, you still need to consider testing from the IT architecture level and involve testers in the early solution phase and throughout the development lifecycle.

An architectural validation test can be based on expert expertise, rules of thumb, spreadsheets, simulation, performance workbench, and so on. *Pre-test architectural efforts* such as static analysis, analogy, and model justification can eliminate potential issues that would otherwise be costly. Table 6-37 shows an A-ESA consideration of the testing process.

Table 6-37. *Testing Process Mapping in A-ESA*

Test Process	Description	A-ESA Notation
Test plan	Risk and issue analysis and considerations	– Risk [RK] – Metrics View [MTS]
	Test method adoption	– Key Choice [KC]
	Test team	– Role [RO]
	Test case	– Use Case [UC] – Document File [DF]
	Seed data test estimate	– Data Service [DS]
	Test tools	– Middleware [MW]
	Control and implementation	– Governance [GV]
Test control	Test monitoring	– Middleware [MW]
	Simulation test	– Operational Walkthrough [VLD]
	Test governance	– Governance function [GV-FN]
Test implementation	Compatibility test	– DevTestOps [DEV]
	Performance and volume test	
	Security test	
	A/B testing	
	Integration test	
	Stress/load test	

Notes:
- Depending on the solution nature, there are many more tests such as usability test, smoke test, cloud test, regression test, user acceptance testing (UAT), etc.
- Depending on your application testing approach (test-driven design or behavior-driven design), your testing process will vary slightly.
- API is a good starting point for testing NFRs.
- A test plan is a must for the A-ESA model, and test control and execution is part of the governance process.

Most tests are for validation. However, A-ESA should be both *validatable* and *viable*. Solution assurance (validated solution) is not the same as long-term architectural conformance. A-ESA governance goes beyond simple solution testing to architectural viability.

HINT

Testing Trends: Emerging trends in testing have a significant impact on ESA, such as AI-driven testing, shift-left testing (moving to an earlier stage in the development cycle), AI-assisted frameworks and tools (Applitools, Testsigma, Katalon Studio, etc.), test data management, predictive testing, and cross-skill quality culture.

Risk Management

ESA is constrained by the solution context, and it requires transparency about potential risks and risk management. Table 6-38 shows the risk mediating elements in A-ESA.

Table 6-38. *Risk Management Elements in A-ESA*

Category	Description	A-ESA Notation
Potential issues	Risk classification and identification	Risk [RK]
	Risk monitoring	Middleware [MW]
Assessment	Initial risk assessment and residual risk assessment	Key Choice [KC]
Containment	Gap clarification and risk mitigation	Governance [GV] Validation [VLD]
Notes: — It's essential to perform root cause analysis (RCA) and correlate the potential issues. — Risk mitigation must be actionable, such as a specific rollback strategy for failed deployments.		

Risk management is part of solution architecture and solution management. A-ESA requires a good understanding of risk management and risk mitigation, so the risk element is essential in an enterprise solution architecture.

HINT

Risk vs. Constraint: For simplicity, A-ESA defines a risk element for a risk, issue, or constraint. Risk has to do with the probability of failure and is closely related to constraints. Constraints can increase or decrease risk, depending on their specific nature. Some constraints may limit the team's ability to perform, while others may actually enhance that ability.

Asset Management

The A-ESA model itself is an asset, based on the existing assets, that can be used for future solutions. A good, large solution model always uses many solution RAs, patterns, or assets. Table 6-39 shows how A-ESA embodies assets when modeling. Note that A-ESA advocates succinct, proven, and graphical assets for easy reference and practical reuse.

Table 6-39. *Asset Management Elements in A-ESA*

Asset Category	Description	A-ESA
Reference architecture	Reference architecture or standard models	– RA View [PTN]
Patterns	Proven patterns from difference sources	– Pattern View [PTN]
Capability repository	Relatively stable and categorized enterprise and solution capabilities	– Capability [CP] – Repository [RP]
Service catalog	Service or service interface element list in the enterprise	– Service Interface [SI] – Repository [RP]
Principle or governance catalog	Principle or governance element list in the enterprise	– Principle [PR] – Governance [GV]

(continued)

437

Table 6-39. (*continued*)

Asset Category	Description	A-ESA
Architecture decision record	Key considerations made for enterprise solutions	– Key Choice [KC] – Repository [RP]
Architectural debt	Architectural and technical debt backlog	– Risk [RK] – Repository [RP]
Solution plateau	Staging deliverable or important solution piece	– View Frame [VF]
Design guidance	Architectural guidance for solution design implementation	– Document File [DF]
Solution repository	Solution model	– Model [ML]
Note: – Assets come in many forms. They can be detailed as artifact [AR] or document file [DF] in the form of libraries, inventories, reports, frameworks, standards, conventions, best practices, and lessons learned. The ESA architect determines their use, customization, and lifecycle management.		

Note that every A-ESA model is a potential asset unless it is scratchable. Although version control is not recommended for A-ESA models, an A-ESA snapshot must be saved and backed up for any major change.

Guide 8: Architecture Adoption Event (AAE)

Last but not least is the architecture method landing guide. A-ESA modeling is not a silent process. It requires an awareness or mindset change on the part of solution teams. An Architecture Adoption Event (AAE) must be held regularly to ensure that modeling is effective. Because architecture adoption is an evolutionary process, it's a series of events.

AAE usually includes a Model Adoption or Model Adaptation Workshop (MAW) as part of the exercises, since customization reflects the way the architecture will be constructed. MAW is a must for A-ESA, and AAE is a must for a large solution with many people, teams, and architectural needs.

AAE reexamines architectural and modeling practices (including gains, pains, and skills) and receives feedback for timely improvement.

Note that the AAE is not about architectural governance or validation per se, nor about any specific architectural tradeoff decision. It's about how to properly represent the ESA architecture, how to reflect architectural thinking in the solution model, and how to improve modeling skills. It may also involve discussions about the overall effect of a chosen architectural style. An AAE serves as architectural management to guide the modeling work in the right direction.

AAE circles around six constituents of this book and answers the questions listed in Table 6-40.

Table 6-40. *AAE Coverage*

Coverage	Sample Questions
Architectural thinking framework	– Does the architecture or model reflect holistic thinking or the architectural thinking framework? – Is the ESA overthought or too complicated? Is the solution architecture worth the current effort? – Does the solution move the right way? – What are the major causes of technical or architectural debt?
Governance technique	– Which governance techniques are most useful for the solution? – How do you apply a particular technique, such as governance functions knowhow? – Does the tradeoff decision-making process fit the solution environment?
NFR or QAR	– Are the metric definitions good enough? – How are these metrics managed (who, which tool, effectiveness)? – Since A-ESA is about long-term sustainability, how are future cases considered?
Architectural style	– Why is the architectural style of the solution? – What is the evolution path or improvements? – What types of architectural elements are included in the style? – How does the chosen architecture style affect the complexity, scalability, and cost-effectiveness of the solution?
Modeling skill	– How competent is the solution architecture and modeling? What skills are required? – Is there a modeling tool? – What improvements can be made to construct solution architecture?
Model specification	– What is the best use of the model specification? – Do you follow a lean mode, a standard mode, or a multiple mode approach? – Which form of model will be used in which case?

Note:
– AAE is an event with a flexible meeting style. It can be described in other ways if necessary.
– It is recommended that an ESA guru be invited for initial guidance if necessary.

For the AAE to be effective, an ESA architect must set up a basic event environment or provide basic training:

- Start with the lean set of A-ESA notations for an initial architecture.

- Create or use an initial skeleton model to gain consensus (on representative notations, etc.) among model users and key stakeholders.

- Clarify the solution process and environment and better align the model with solution management.

- Assign responsibilities to each model stakeholder, including an owner role, and assess interest and qualifications.

- Ensure that all team members understand A-ESA relevance and proliferation, with typical examples to illustrate.

In short, an ESA reexamination or retrospective ensures an overall consideration of governance techniques. The key is to guide the ESA in the *right way*, not just to do the *right thing*, which is solution-assured architecture or modeling work.

HINT

Architectural Thinking vs. Overthinking: IT architecture simplifies and clarifies complexity. While A-ESA takes a holistic approach, keep simplicity in mind. Avoid overthinking by adopting a "divide and conquer" and iteration approach. Many architectural problems can be solved from a multi-dimensional viewpoint.

Summary

Governance is often overlooked in many enterprise solutions. It is sometimes an afterthought. As one of the key constituents in A-ESA, governance ensures the viability of the ESA. This chapter has provided various governance and techniques for an enterprise solution architecture. Applying proven architecture techniques is an important part of governance. Techniques and governance are complementary in an ESA.

This chapter also discussed the relevance of solution management, which is part of the ESA context. Solution architecture is the core player and solution management is the supporting partner.

This chapter concludes the book with governance and techniques. Techniques are easy to learn, but hard to apply well without deep thinking and hands-on modeling practice. Each governance technique is applied with different rules for distinctive architectural styles and different sets of non-functional requirements. Thus, A-ESA architects must consider all six interdependent constituents described at the beginning of the book to produce an effective architecture. An AAE can provide valuable guidance and foresight to ensure that the modeling work is conducted in an optimal manner.

Epilogue

Over the past 30 years, I have held various IT positions in different industries: ITSP consultant, CTO, enterprise architect, ERP architect, business analyst, software engineer in multiple programming languages, solution architect (integration, MSA, cloud, etc.), data modeler, Level 3 IT support and operational manager, and so on. My myriad of large, hands-on IT project experience as a chief/lead architect shows that A-ESA is critical to building a sound organization-wide information systems architecture. An enterprise solution architecture without vetted modeling will result in either a sacrificial architecture or a lousy deliverable. I hope that this book can better contribute to the world of enterprise solution architecture.

If you imagine the future, enterprise solution architecture will be supported by twin drivers: IT hosting (cloud service or similar) and AI computerization. In spite of all the changes in IT, the fundamental ESA is still the same.

Public hosting services are growing at a rapid pace, moving much of the IT infrastructure and middleware capabilities to the hosting platform. This trend will replace much of the technical architecture work, part of the data and application architecture. However, for enterprise solution architecture, the seemingly heavy technical architecture itself is always the easiest part, as the application logic and distributed data integrity consume much of the ESA work, and business agility will become more necessary over time. Ultimately, whether third-party hosting, homegrown, or hybrid, ESA still addresses many NFR considerations, including application flexibility, cost-effectiveness, ownership, and security.

© Sean (Chunhong) Gu 2024
S. Gu, *Mastering Enterprise Solution Modeling*,
https://doi.org/10.1007/979-8-8688-0992-7

Similarly, *AI-driven automation* will replace much of the mechanical work, such as common algorithms, template-style coding, and technical scalability. It will certainly automate much of the validation work of the architecture tools and simplify what-if simulations, with real-time visual rendering of architectural information, operational walkthroughs, parametric configuration, and proactive problem detection. However, the *tough or art part* of enterprise solution architecture, which requires intuitive requirement mapping and architectural thinking, will remain in the minds of architects. This makes architects *one of the least threatened professions* in the AI era. Powered by AI, ESA architects will be better and faster at making judgments and building consensus among diverse stakeholders. It is human reasoning and the ability to think outside the box and consider unforeseen circumstances that is essential to holistic modeling. In short, AI will help architects truly reach the upper echelon of less validation, more critical thinking. Likewise, the practical, robust A-ESA modeling approach will facilitate AI enablement for greater insight and wider adoption. For ESA, AI is designed to help, not to rely on. AI-assisted A-ESA is expected to prevail.

Thanks for your interest in ESA/A-ESA. I can be reached at guc888@gmail.com or contact@a-esa.com for comments, criticism, consultation requests, and tool support.

APPENDIX I

A-ESA Spec Addendum

A-ESA Base Element List

Table I-1 is an A-ESA base element list. The iconic notations follow the familiar specifications found in many enterprise and solution architectures.

© Sean (Chunhong) Gu 2024
S. Gu, *Mastering Enterprise Solution Modeling*,
https://doi.org/10.1007/979-8-8688-0992-7

Table I-1. *A-ESA Element List*

Area/Category	Element	Abbr.	Icon
Enterprise	Capability	CP	
	Value/Value Stream	VL	
Case Scenario	Role/Actor	RO	
	Task (Activity, Operation)	TK	
	Use Case/User Story	UC	
Metrics	Principle/Guidance	PR	
	Requirement/Rule	RQ	
	Key Choice/Consideration	KC	
	Risk/Constraint/Arch Debt	RK	
	Governance	GV	
Functional Service	UI (User Interface/Interaction) Service	UI	
	App Logic Service	AS	
	Data Service	DS	
	Tech Service	TS	
	Service Interface	SI	
	Service Component	SC	

(continued)

Table I-1. (*continued*)

Area/Category	Element	Abbr.	Icon
Operational Service	Deployment Package/Unit	DP	
	Middleware	MW	
	System/Device	SY	
	Node	ND	
	Network	NW	
	Location	LO	
Connection	Association	AN	
	Flow	FW	
	Composition	CN	
	Realization	RN	

(*continued*)

Table I-1. (*continued*)

Area/Category	Element	Abbr.	Icon
General	Note	NT	
	Grouping	GP	
	Generic Domain	DM	
	Generic Service	GS	
	Virtual Service	VS	
	View Frame	VF	
	Deliverable	DL	
	Artifact	AR	
	Extension	EX	

Note:

— Some elements, such as generic services, middleware, and extensions, often have customized images for distinctive visual representations.

— A-ESA emphasizes significant requirement cases, so the case requirement elements (such as UC and RQ) denote key or typical cases.

— Dotted or dashed element borderlines can be used to indicate virtual or logical elements, and dotted connection lines indicate asynchronous or interruptible communication.

A-ESA Base Element Definition

Table I-2 is a list of short definitions for the A-ESA base elements.

Table I-2. *Definitions of Base Elements*

Element	Definition
Capability	Represents an EA-level ability that a structural element (such as a functional service) possesses.
Value	Represents the relative importance, worth of a concept, or vision from an enterprise perspective.
Role	A responsibility for performing a specific behavior. It represents an actor, a user, or a user group.
Task	A business function, activity, or piece of work assigned to a role in a process.
Use Case	Describes the interactions between a role and a system to achieve a goal.
Architecture Principle	Represents a qualitative statement of intent that must be met by the architecture. It's part of a guideline framework.
Requirement	Represents a concise statement of needs in a broad sense.
Key Choice	An architecturally significant decision, gap analysis, assessment, issue resolution, or solution assurance based on the collective solution strategy.
Risk	Represents a constraint, a potential issue, something that hasn't happened, or key decisions that need to be addressed.
Governance	Embodies IT architecture standards, compliance, and criteria.
UI Service	An interactive service usually with a visual presentation.
Application Logic Service	Represents explicitly defined non-GUI application behavior and control logic, or composition.

(continued)

Table I-2. (*continued*)

Element	Definition
Data Service	A self-contained piece of information that has a clear meaning to the application, and can be a standalone data object, or a federated, integrated piece of data service.
Technical Service	Represents a behavior that is independent of the application-specific logic context.
Service Interface	Represents an access point where services are made available to a user, service, or service component.
Service Component	Represents the implementation of service responsibility or functionality.
Deployment Package	Contains functional services based on their unique service-level characteristics.
Middleware	Represents system or intermediate software that provides services to software applications.
System Device	A collection of hardware and software pieces and a set of relationships for specific business functions.
Node	Generally represents a hosting resource that interacts with other resources.
Network	Denotes a set of structures, products, and services that enable the connection of system nodes for data transmission.
Location	A place where structural elements are positioned or communicated.
Association	Represents a generic or an unspecified relationship.
Flow	Represents the movement from one element to another.

(*continued*)

Table I-2. (continued)

Element	Definition
Composition	Composed of one or more other elements.
Realization	Moves from an abstract element to a concrete one.
Note	Represents the commentary or interpretation of the architecture.
Group	Represents a logical representation, a layer, or a generic composition or aggregation of elements.
Domain	A concept of the functionality or boundary where multiple elements are controlled under the same scope.
Generic Service	Covers well-defined business or application activities in a specific solution context and serves as a logical service, an architectural service, or a functional service.
Virtual Service	Represents an intangible asset or information, or a technical service without a physical IT service form or clear interface.
View Frame	Used to help scale an architectural design as a drilldown view or solution plateau.
Deliverable	Represents a defined outcome, generally more of a concern from a solution management perspective.
Artifact	Represents an additional solution specification or design piece. It can be a class, method, namespace, document, project, or the like.
Extension	A flexible representation for model element addition, including a non-IT element or external system element.

A-ESA Assistive Element List

As an Agile model, assistive elements can be added to the base element list and listed as sub-elements. Table I-3 shows some common assistive elements used in an enterprise solution environment for your reference.

Table I-3. *A-ESA Assistive Element List*

Base Element	Assistive Element	Prefix	Usage (H/M/L)
Artifact	Document File	DF	H/M
	Entity	EN	M
	Object	OB	L
	Table	TB	M
	Repository (Library, Catalog)	RP	M
	Model	ML	M
	Domain Specific Language (DSL)	SL	L
	Arch Decision Record	AD	H
	Sample Data	SD	H
	Tabular Data Info	TD	H
	Item(s)	II	H
Deployment Package	Microservice	MS	H/M
	Runtime Application	RA	L

(continued)

Table I-3. (*continued*)

Base Element	Assistive Element	Prefix	Usage (H/M/L)
Domain	Module	MO	M
	Schema	SM	M
	Bounded Context	BC	H/M
	Transaction	TR	H/M
	Cluster	CT	M
	Partition	PT	M/L
	Cell	CE	L
	Zone	ZN	M
	Platform	PL	M
	Environment	EV	M
	Tier	TE	M/L
Location	Region	RG	M
	Cloud (External Provider, External Dependency)	CL	H
Data Service	Message	MG	H/M
	Cache	CA	H/M
	Session	SS	L
Generic Service	Process/Composite	PS	H
	Application	AP	M
	Domain Service [1]	DM	H/M
	Product	PD	L
	Event Service	ES	M

(*continued*)

Table I-3. (*continued*)

Base Element	Assistive Element	Prefix	Usage (H/M/L)
Governance	Governance Function	GF	H
Middleware	Database	DB	H
	Edge Interface	EG	H
	Message Broker	MB	H
	Event Broker	EB	M
Extension	Equipment/Rack	EE	L
	Materials	MA	L
Requirement	Goal	GL	L
	Non-functional Requirement	NF	H
	Rule	RL	M
Service Interface	Service/Component API	PI	H
System Device	Mobile Device	MD	M
	PC Device	PC	M
Tech Service	Generic Implementation	IM	M
	Message Queue	MQ	H
	Event Queue	EQ	M
User Interface	Browser	BR	M
	Frontend Client	FE	H/M
Service Component	Function (FaaS)	FC	M

(*continued*)

Table I-3. (*continued*)

Base Element	Assistive Element	Prefix	Usage (H/M/L)
Virtual Service	Portlet/Widget Collaboration	VC	L
	Config File	CF	M
	Virtual Node	VN	M
Grouping	Resource Group	RS	M
Directional Relationship	Signal	SG	L
	Trigger	TG	M

Note:

- [1] In A-ESA, domain is more of an application or technical boundary concept. It's up to you to use domain element as a problem space boundary. The key is to clarify its true intent or result in an A-ESA context.
- Mapping the assistive element to its base element depends on the solution context. For example, the *cloud* element can be mapped to a *location* or *domain* element.
- The prefix of an assistant element can also be expressed as the prefix of the base element plus the prefix of the assistive element, for example, document file prefix = [AR-DF].
- The Item element is meant to visually express exemplary element content in general.
- You can use a familiar notation to better fit your communication culture. For example, you can use the generic service as a business service (GS-BS), even though A-ESA does not include the fuzzy business service concept.

A-ESA Commonly Used Elements

A-ESA elements are customizable to meet the needs of each solution. Each architecture style has its own preferred element usage. For example, an application architecture obviously requires a different set of assistive elements than an integration architecture. In addition, some assistive elements (such as MS, FE, EG, DB, CL, MQ) are needed more frequently than others.

In general, base elements convey the essence of the enterprise solution architecture, with assistive element or special image notation for better visual and customized representation. Chapter 2 demonstrates A-ESA modeling using base elements, assistive elements, and special images. Table I-4 shows the different element usage scope for each solution profile.

Table I-4. *A-ESA Element Usage Scope*

Profile No.	Element Scope	Explanation
1, 2, 8	Base Element	– Profile 1 used the Cloud [CL] element – Profile 2 used the Microservice [MS] element
3, 5, 6	Base Element + Assistive Element	– Application architecture generally requires some assistive elements for an end-to-end scenario walkthrough.
4, 7	Base Element + Assistive Element + Special Image	– Profile 4 used custom graphics to represent some specific Middleware [MW] and Generic Service [GS] in the solution. – Profile 7 used vendor's images.

Table I-5 shows a list of the assistive element images/icons used in the solution profiles in Chapter 2. Although they appear as different notations, they still fall within the A-ESA basic element group.

Table I-5. *Assistive Elements used in Chapter 2*

Prefix	Assistive Element	Icon	A-ESA Category
BC	Bounded Context		Domain
BR	Browser		UI (User Interaction)
CA	Cache		Data Service
CE	Cell		Domain
CT	Cluster		Domain
DF	Document File	.xyz	Artifact
EG	Edge Interface		Technical Service
EN	Entity		Artifact
EV	Environment		Domain
FE	Frontend		UI (User Interaction)
GF	Governance Function	Fn	Governance
MD	Mobile Device		System/Device
MQ	Message Queue		Technical Service

(continued)

Table I-5. (*continued*)

Prefix	Assistive Element	Icon	A-ESA Category
NF	Non-functional Requirement		Requirement
PL	Platform		Domain
PS	Process		Generic Service
RG	Region		Domain/Location
RL	Rule		Requirement
SM	Data Schema		Domain
ST	Stream		Data Service
TB	Data Table		Artifact
TR	Transaction		Domain
ZN	Zone		Domain/Location

As you can see from the eight modeling examples, base/basic elements dominate. Each solution may require a few assistive elements or special images for a full expression of the solution architecture. A rule of thumb is about 40 types of elements (base and assistive elements) for a typical enterprise solution.

The A-ESA model specification is based on numerous projects, and it fits most enterprise solution needs. Limiting the assistive elements within the base element scope ensures the simplicity of the A-ESA model.

A-ESA Commonly Used View Areas

A-ESA has a list of recommended architecture areas and views (see Table 1-1). However, it is Agile to specify your own view areas for solution needs. Table I-6 shows common A-ESA area/view coverage. Although areas or views may be defined differently, they fall under this table definition.

Table I-6. *Commonly Used View Areas*

Area	View	Common Coverage
Enterprise	Capability	Capability and resource map
	Organization	Role and stakeholders
Case Scenario	Process	Business process and value stream
	Use Case Model	Requirement mapping
Architecture Overview	Outline	Integration, migration, layering, and division of work
	Metrics	Strategy and driver/goal
	Pattern	Reference architecture, assets, and guidance
	DevOps	Implementation and deployment
Functional	Functional Relationship	Application structure, information structure, and relationship
	Interaction	Application interaction and collaboration
	Realization	Service realization and outcome realization
Operational	Operational	Technology usage and physical environment

A-ESA Naming Conventions

Naming is part of the A-ESA modeling process. It is flexible but must clearly reflect its A-ESA relevance. Table I-7 provides some common naming conventions.

Table I-7. *A-ESA Naming Convention Examples*

Notation	Naming Convention	Example
View	View Prefix + ID	PTN-1 (Architecture Pattern, ID=1)
	Area Prefix + View Prefix + ID	AOV-PTN-1 (Architecture Overview, Pattern, ID=1)
Element	Base Element Prefix + ID	RQ-1 (Requirement, ID=1)
	Base Element Prefix + Subcategory Prefix + ID	RQ-NFR-1 (Requirement, Non-functional Requirement, ID=1)
	Subcategory Prefix + ID	NF-1 (Non-functional Requirement, ID=1)
Note:		
— The prefix or ID can be hidden from view to accommodate display spacing.		

APPENDIX II

A-ESA Tools

A-ESA is a modeling approach. It also comes with some home-grown tools to support a flexible and robust modeling approach.

All figures in this book are from a customizable A-ESA tool that uses a list of predefined views (see Table 1-1 in Chapter 1), standard elements (see Table 1-3 in Chapter 1), and assistive elements (see Table I-3), with a set of iconic notations and preferred display styles. Several points to note here:

- In general, it is advisable to use node-style shapes rather than box-like shapes. This is subtle but visually effective for A-ESA.

- A-ESA is meant to explicitly convey architectural intent, and modeling forces you to think concisely and holistically.

- A valid A-ESA architecture needs to be embodied by the simplicity, significance, and systematics of the model tool. Much of the verification or validation work can be automated.

- Finally, you are encouraged to model A-ESA using the lean mode.

© Sean (Chunhong) Gu 2024
S. Gu, *Mastering Enterprise Solution Modeling*,
https://doi.org/10.1007/979-8-8688-0992-7

Without modeling support, large-scale architectural thinking is limited. To err is human. Tools help streamline and correct obvious bugaboos such as duplication and inconsistencies. The modeling examples in the book only touch on basic A-ESA techniques. The model presents key A-ESA architectural descriptions, diagrammatic representations, correlation, and verification. There is more to be covered in terms of tool support, such as walkthrough simulations, algorithmic automation, and artificial intelligence. With AI support, the A-ESA tool will become more constructive and powerful.

A-ESA prefers to present a correlated set of model notations for easy automated verification and validation. However, the sketch diagram has its own use for quick and easy modeling in certain cases. The A-ESA tool will also support diagram views for Agile users, as many architects now do, similar to the uncorrelated Visio views.

Note that the uniqueness of the A-ESA tool is not about automation or sophisticated features, but rather about improving architectural thinking, as outlined in the A-ESA thinking framework.

The A-ESA tool is initially available in a basic version, and advanced versions will be available later after feedback. Whatever the tool, the goal is to keep it simple. If you are interested in the tool, visit a-esa.com for more information.

APPENDIX III

Modeling Language Comparison

A-ESA is an architectural approach and a modeling language. Its scope can be well understood by comparing it to some of the popular modeling specifications, as listed in Table III-1.

Table III-1. *Comparison with Some Popular Languages*

Popular Specs	Coverage Comparison
ArchiMate	ArchiMate is an enterprise architecture specification, and it covers more at the enterprise level, while A-ESA is IT-service-based and covers more on the transition from enterprise architecture to solution architecture. It focuses on solution building blocks, non-functional characteristics, and key architecture decisions. For example, more domain related elements considering transaction and partition, etc.
SoaML	SoaML is a specification for SOA architectural design and solution integration, focusing on service communication, while A-ESA identifies and specifies IT services from business architecture, application architecture, data architecture, and technical architecture in a holistic approach and covers more enterprise concerns.

(continued)

© Sean (Chunhong) Gu 2024
S. Gu, *Mastering Enterprise Solution Modeling*,
https://doi.org/10.1007/979-8-8688-0992-7

Table III-1. (*continued*)

Popular Specs	Coverage Comparison
SysML	SysML is mainly for system engineering architectural design, and an MBSE[1] specification with nine types of diagrams, while A-ESA is a relatively high-level system solution architecture with less on system engineering.
UML	UML is primarily aimed at software architecture and design, or software engineering delving into classes and objects, although it contains a node notation and can be used as a low-level solution architecture. A-ESA is a system solution architecture and provides a blueprint for further software designs.

Note:
— Broadly speaking, these modeling languages can also be referred to as methodologies, specifications, and frameworks.

As you can see from Table III-1, each architectural design specification has its own purpose and use. By design, A-ESA strikes a delicate balance between EA tools (ArchiMate) and the architectural design tools (UML, etc.).

For software solution architecture, there are many modeling approaches, simple and complex. For example, the simple C4 model approach can be used for solution visualization based on a set of hierarchical abstractions (software systems, containers, components, and code), which is notation independent and represents a different level of abstraction. However, such a model often needs to incorporate other models (such as UML/BPMN) as well as metric elements or additional notations when considering A-ESA-like basic or standard elements for correlated, systematic architectural conformance.

In short, whatever approaches you take to solution architecture, remember that an architecture is, by definition, NOT just about structures and relationships. A-ESA is designed to provide a solution architectural guidance model without relying on prevailing modeling tools.

---

[1] Model-Based Systems Engineering

APPENDIX IV

ESA Architect's Roles and Skills

An ESA architect serves as a liaison between the strategic planning and design and implementation of an enterprise IT, and importantly, builds a working model that guides and reflects the real solution system.

The ESA architect role is not limited to an architect job title. In an Agile architecture environment, an ESA architect role can be filled by an aspiring developer who is good at problem solving, a designer who is good at systems thinking, or a team member who is good at architectural thinking. In short, an ESA architect should:

- Have sufficient and broad hands-on experience and specialization in at least one A-ESA domain (case scenario, functional, or operational), fluency in at least one architectural style (microservice, cloud architecture), with T-shaped business domain knowledge and technical expertise.

- Critically, have a good grasp of appropriate abstraction skills and not delve into architectural design details.

- Be good at issue resolution and trade off balancing between different service-level features.

© Sean (Chunhong) Gu 2024
S. Gu, *Mastering Enterprise Solution Modeling*,
https://doi.org/10.1007/979-8-8688-0992-7

- Be able to build a realizable model that includes people, processes, and technologies by specifying governance functions and exemplifying significant concerns via realizable feature attributes, simulations, and prototypes (POT[2] or POC).

- Be able to accumulate reusable assets (templates, patterns, guides, and automation tools).

From an enterprise architecture perspective, the focal skill requirements for solution architects are assessed in Table IV-1.

Table IV-1. *Enterprise Solution Architecture Skills*

Roles	Enterprise Solution Architect					Solution Architect
	Chief/Lead	Business	Application	Data	Technology	
Leadership Capability	4	3	3	3	3	2
IT Industry Standard	4	4	4	4	4	3
Benefits Analysis	4	4	4	4	4	3
Business Scenario	4	4	4	4	4	4
Business Modeling	4	4	4	3	3	2
Architecture Principles	4	4	4	4	4	3

(*continued*)

---

[2] Proof of Technology

Table IV-1. (continued)

Roles	Enterprise Solution Architect					Solution Architect
	Chief/Lead	Business	Application	Data	Technology	
Views & Viewpoints	4	4	4	4	4	4
Building Block Design	4	4	4	4	4	3
Modeling Skills	4	4	4	4	4	4
User Interface	3	3	4	4	4	3
Programming Languages	3	3	3	3	3	4
Software Engineering	3	2	4	4	3	4
Application Deployment	3	1	4	3	3	3
Application Integration	4	3	4	4	3	4
Service Design	4	3	4	4	4	3
Data Interchange	3	2	3	4	4	3
Transaction Processing	3	3	4	4	4	3
System Integration	4	3	3	3	4	3

(continued)

Table IV-1. (*continued*)

Roles	Enterprise Solution Architect					Solution Architect
	Chief/Lead	Business	Application	Data	Technology	
Computing Infrastructure	3	2	3	3	4	3
Migration Planning	4	3	3	4	3	4
Service Level Agreements	4	3	4	3	4	3
End-user Support	1	1	3	2	2	2
Project Management	3	3	3	3	3	2
Program Management	3	3	3	3	3	1
Legal Protection	4	3	3	4	3	2

Note:
- Proficiency level: 4-Expert, 3-Knowledge, 2-Awareness, and 1-Background
- The key criterion to evaluate a solution architect are modeling skills.
- Source: Adapted from "Architecture Skills Framework," The Open Group, 2022.

The qualifications of an enterprise solution architect can be referenced through the Open Certified Architect (Open CA), which offers three levels of architect certification:

- **Level 1:** A professional capable of working with assistance/supervision and contribute a broad range of appropriate skills.

- **Level 2:** A professional capable of performing independently and taking responsibility for the delivery of solutions as a lead.

- **Level 3:** A professional who has a significant breadth and depth of impact on the business through the application of their profession.

Note that an ESA architect is *not an armchair general*, but rather a hands-on practitioner of viable solutions. The central skill for an ESA architect is the ability to model, which is a unique skillset that differs from many other professional roles. Modeling fosters systematic thinking. A PPT or paper architect will never be a qualified ESA architect. The other implicit skill for an ESA architect is the ability to manage complexity. ESA is about simplifying complex architectural problems, not just problem identification, analysis, or lengthy documentation.

In short, in addition to modeling skills, an ESA architect is usually customer-facing and requires *broad views, deep and versatile expertise* with troubleshooting and quick learning skills. After all, an IT architecture is a *multidimensional thinker* with *rich experience*. Therefore, the ESA role fits only a select group of IT professionals.

APPENDIX V

FAQs about A-ESA

There are some Frequently Asked Questions (FAQs) that have been asked of me about Agile Enterprise Solution Architecture (A-ESA). Although the book provides answers to these questions one way or another, this appendix provides brief recaps for your convenience. See a-esa.com for more FAQs.

About A-ESA's Relevance to Other Architectures

The following are FAQs about A-ESA's relevance to other architectures.

How Does ESA Differ from EA and SA?

In fact, there is no standard definition of Enterprise Architecture (EA), which is a comprehensive framework that defines the structure of an organization and explains how its various components work together to achieve its goals and objectives. Essentially, it's a blueprint that outlines how an organization should operate and leverage its resources to be efficient, effective, and adaptable to change. EA typically encompasses several domains, such as business architecture, data architecture, application architecture, and technology architecture. *It's a strategic tool.* EA scenarios include digital transformation, business target operating model, overall enterprise landscape, and so on. EA can be ESA if it exhibits a similar framework or model as mentioned in this book.

© Sean (Chunhong) Gu 2024
S. Gu, *Mastering Enterprise Solution Modeling*,
https://doi.org/10.1007/979-8-8688-0992-7

Enterprise Solution Architecture (ESA) is more closely aligned with the IT strategic plan (ITSP) and leans toward solution architecture (SA). It maps the strategic intent of EA and creates a solution blueprint from an enterprise perspective. *It's the link between EA and SA.*

SA is a comprehensive approach to defining the structural components, technical strategies, and guidelines. It is generally a more detailed architecture in a specific area or small scope and meets the requirements of a project or business need. When an SA is large and complex (commonly seen in a multi-project program), it can be ESA in nature. Or SA can be so detail-oriented that it's practically a solution design.

By way of comparison, the following analogy can be drawn between EA, ESA, and SA: EA can be compared to urban planning, SA to a building model, and ESA to a multi-village community masterplan model that produces a balance between the desirability of the outcome and the environmental impact.

What's the Enterprise Systems Architecture?

This book is about enterprise solution architecture, the link between EA and SA, as mentioned earlier.

Enterprise systems architecture is part of EA. It focuses on technology to support business. In some cases, enterprise systems architecture is the same as technology architecture, especially when the focus is on the technology and how it works within the company. In other cases, enterprise systems architecture is like technology architecture, but it also addresses how systems align with business goals and optimize business processes. Sometimes ESA (enterprise systems architecture) and EA are interchangeable terms and have more landing considerations.

When enterprise systems architecture is an extended version, it overlaps significantly with enterprise solution architecture. Enterprise solution architecture involves software engineering and systems engineering to design and implement solutions that meet business needs within the context of the enterprise systems infrastructure. While enterprise systems architecture focuses primarily on systems engineering, it may also include software engineering to the extent that software is a component of the systems infrastructure being designed and managed.

In practice, the architecture of a solution should always be considered in the context of the larger enterprise systems, even if it is not enterprise-wide. This ensures that the solution can be effectively integrated with existing systems and contribute to the organization's overall goals and objectives.

Simply put, Enterprise Systems Architecture and Enterprise Solution Architecture may be used interchangeably, if they have the same coverage and accomplish the same goal. Therefore, the naming is one thing, and the real intent is another.

Incidentally, other terms are often used interchangeably, such as solution architecture vs. solutions architecture, and system architecture vs. systems architecture, although the use of the plural form can sometimes indicate a broader scope or a more collective view of the architectures being discussed. However, in most contexts, solution architecture and solutions architecture mean the same thing.

Why Is ESA More Demanding than EA and SA?

Traditionally, EA people are strategic thinkers or high-level planners. SA folks are more technical and specialized. However, ESA architects need mindsets and expertise from both ends of the spectrum: EA and SA. On the one hand, fewer people have had the opportunity to work in both ends; on the other hand, it requires diverse domain knowhow, up-to-date IT knowledge, and tremendous effort. *The combination of broad skill requirements and tough job role responsibilities is challenging.* In addition, business leaders tend to understand only EA to the neglect of SA. Some people believe that if they have a well-planned EA, magic will happen in SA. In some organizations, an ESA architect is still an obscure role.

The IT industry has not had a common language for ESA, and there is no standard or organization dedicated to ESA, unlike EA, which is sponsored and maintained by The Open Group. From a tooling perspective, there is currently no well-defined ESA tool on the market. ESA architects often use both EA and solution design tools, which are difficult to correlate in a large and complex solution.

A significant role of ESA is to address the phenomenon of emergent behavior, which encompasses the attributes, patterns, and occurrences that emerge from the interaction of individual components within a system. The emergence of these behaviors may not be readily discernible in a single or limited solution, nor in a comprehensive EA model. While AI can facilitate prediction, it must be founded on a robust ESA.

What Factors Contribute to EA's Ineffectiveness in Many Cases?

From an IT architecture standpoint, many EA projects, proposals, blueprints, and plans place an undue emphasis on frameworks, processes, and methodologies without adequately accounting for ongoing changes in the solution environment. Some EA reports and presentations appear sophisticated and theoretically sound, but *are not aligned with the practical realities and priorities of the solutions being developed.* In contrast, ESA integrates a systematic engineering process to a certain extent, to ensure that the quality of the architecture meets the business needs.

It is important to note that EA still serves as a strategic planning or roadmap function. When it considers solutions, as ESA does, it can also guide the development of those solutions.

Why Is the Traditional ITA Not Working Well?

Traditional ITA (IT Architecture) works in a one-way approach to guiding the solution architectural design. Such an architecture can seem pretty lofty, and it doesn't always connect with the development team. Their feedback may not make it back to the architecture.

A-ESA fosters teamwork and collaboration. The ESA architect is not a solitary decision maker, but rather a role that draws wisdom from the entire architecture process. As the A-ESA name suggests, it takes an Agile approach to iterating the architecture and walking through significant cases. A-ESA reflects not only the strategic intent of the enterprise, but also the solution realization approach and techniques. It facilitates the bi-directional flow of business solution needs, solution architecture decisions, and solution development, and it helps connect the two ends of the solution architecture.

In short, A-ESA's down-to-earth, end-to-end approach helps break down the traditional barriers between EA and SA, between SA and SD (solution design or solution development), and between EA and SD.

How Can Software Architects Take Advantage of A-ESA?

For a large or distributed solution, the software architecture is the core reason for maintainability, and therefore flexibility. A software application with many intelligent components and a good design does not mean a good architecture that views the whole solution from a higher level.

Software architecture can take the traditional MVC (model-view-controller) approach and consider the model as a data service, the view as a UI service, and the controller as an application logic service, plus utilities as a reusable technical service. Think in terms of guidance, significant case scenarios, and system architecture to align metrics and map deployment packages. Remember, good solution architecture drives solution design, not the other way around.

What Is the Role of the ESA Model in Enterprise Transformation?

Enterprise transformation is an important part of EA. It commonly provides a clear roadmap for IT transformation. An ESA model addresses specific pain points and requirements and maps dependencies and changes across different parts of the architecture. IT transformation and migration requires a concerted effort. The ESA model takes input from the EA and adopts a phased transformation process:

- Align with heat map analysis in *capability* views and ensure transformation *values*.

- Specify architecture *principles* that clearly reflect business strategies and make the *right choices*.

- Use *view frames* to perform comparative mappings to ensure a clear *AS-IS* and *TO-BE* architecture mapping for *incremental deliverables*, starting with the smallest.

- Work with solution management context, including resources, scope, schedule, and budget using notations like *role, risk, system context, and validation*.

IT transformation projects tend to focus on strategies and processes. However, a clear ESA model supports not only the feasibility reasoning case, but also the landing assurance case.

What's the Scope of ESA Across the Solution Space?

Agile ESA uniquely covers enterprise architecture, software architecture, and system architecture as a whole, but it's not a handyman's tool for all kinds of "solution architectures." It does not target detailed design elements or physical elements beyond IT solutions, as well as operational technology (OT), and many specialized system architectures (MOSA,[3] SOSA,[4] etc.) or solution designs.

Solution architecture and design can cover a wide range. You may find a specialized tool for your solution design needs. For example, AutoCAD for mechanical drawing, EDA (electronic design automation), CircuitLab for circuit design, and so on. You can also use generic tools like Gliffy and draw.io to create informal diagrams for solution architectures and designs.

Clearly, ESA's unique scope and focus sets it apart from other solution architectures. It covers the *enterprise solution architecture* space adequately and broadly. ESA overlaps with many solution architectures to varying degrees.

About the A-ESA Approach

The following are FAQs about the A-ESA Approach.

Why Does A-ESA Take an IT Service-Based Approach?

Software architects and developers tend to design solution architectures based on technical services (or components), leaving little or no room for abstraction, while business people tend to think about business services from a different focus and granularity of technical services.

---

[3] Modular Open Systems Approach
[4] Sensor Open System Architecture

They both look at the same solution from different perspectives. However, ESA architects, who consider both business needs and technical implementation, view the solution from an architectural characterization. ESA uses IT services or architectural services that best reflect the solution architecture. An ESA model includes a well-considered set of IT service notations (functional or operational) and associated notations that define those IT services.

IT service is part of the ESA language, and it strikes a middle ground between business and IT automation. Thus, if well architected, IT services can satisfy business agility and technical feasibility.

Without this layer of abstraction and mapping, a large solution that is subject to change will always be a mess. For example, your solution can still be fragile if you don't have a set of well-architected architectural services, just a bunch of SOA web services or microservices.

How Can I Make the A-ESA Language Easier to Understand?

The A-ESA language is aimed at IT solution architects and has specific connotations for enterprise solutions. How can the A-ESA language play a better role in a collaborative solution environment with people from different backgrounds and skillsets? Here are some A-ESA usage examples:

- **Intuitive ESA elements:** A-ESA model contains a required yet minimal set of views/elements in an IT service-based approach for the right level of abstraction.

- **BizTech style:** A-ESA can be rendered in a more systems architecture approach using functional and operational views whose audience is a professional architect. Or it can be rendered in a business architecture or cartoon-like visual image using capability and case scenario views for a non-IT audience.

- **Lean mode:** A-ESA encourages the use of lean mode for rapid architectural modeling. There may be two different sets of model views between business and IT people, or between the simple view and the standard/specialized view. The ESA architect is responsible for translating between the two.

Who Should Assume Responsibility for ESA?

A-ESA should be assigned an *owner*, which may be a team or a role. Depending on the scope of the solution, the key ESA responsibility may be held by a chief architect for overall enterprise-level issues, a lead architect for a major solution, or a specialist architect for a specific part of the architecture.

ESA includes all stakeholders as participants who provide input or feedback. The ESA architects organize or *map* the input received into a solution with measurable metrics and shape it into the A-ESA model.

What About ESA Tool Automation?

For a large ESA model, tool automation is a must. There are many visual graphical tools, such as Mermaid, that can generate views based on verbal descriptions. AI is now capable of generating basic views for specific solutions or common scenes, and it helps with simulation and validation. While AI excels at complex tasks, it is not infallible. To leverage AI effectively, architects must possess the ability to think critically and make sound judgments.

The prerequisites for automating an ESA model include: 1) a set of ESA-specific, meaningful elements, 2) all elements correlated in multiple views reflecting different model modes and viewpoints, and 3) a composed set of elements that match a specific set of metric elements, which requires a lot of architectural thinking, the art of human wisdom. Everything in ESA is a tradeoff, and balancing this seesaw is the art.

AI ESA involves a series of iterative steps: creating an initial ESA, simulating ESA case scenarios, and making architectural decisions. AI ESA is refined each step along the way. It is a combination of technical automation (AI bot) and AI-assisted reasoning (participating human bot). Key decisions and intuitive adjustments are left to the ESA architects.

About the Agile Architecture

The following are FAQs about the Agile architecture.

Why Is Agile Closely Related to ESA?

Unlike an EA, which has a relatively long-term strategic goal, ESA is a landing plan that is subject to constant change. Agility allows key architectural considerations to be captured dynamically and incrementally in a flexible modeling approach. It reduces the learning curve and leverages common sense to meet different solutions.

Does A-ESA Use Agile Frameworks?

The Agile architecture does not require the use of Agile frameworks such as Scrum and SAFe, although they are a good fit. A-ESA does not require these frameworks either, but advocates their principles. As mentioned in the book, A-ESA does not specify a formal process, but requires a modeling approach that maps critical Agile activities through iterative delivery. Agile architecture and A-ESA both live on the promise of lower change costs, so they are complementary. Ideally, A-ESA is the unity of the Agile methodology and an IT architecture.

Is Design Thinking Valuable for ESA?

Design thinking is a problem-solving methodology that emphasizes user-centeredness, explores problem spaces, and generates innovative solutions through iterative processes and interdisciplinary teamwork. Design thinking helps with requirements gathering and analysis, which is an important input to ESA's case scenario mapping.

ESA emphasizes mapping the problem space to the solution space. A well-defined problem space must be supported by one or more solutions. When design thinking and architectural thinking go hand in hand, there will be a successful ESA outcome.

However, design thinking alone lacks long-term considerations and tends to overlook system issues, technical choices and standards, compliance, governance, accumulated problems, and so on.

About Solution Process and Management

The following are FAQs about solution process and management.

How Does A-ESA Relate to Solution Management?

For ESA, the architecture model is the core, guided by the solution process and supported by solution management. In general, solution process is more important in the initial project phase for design and quality, while during implementation, delivery, and operation, solution management plays a greater role in sustaining solution value. While this book focuses on solution modeling, it also addresses solution process and management, with which a well-designed model should be integrated.

The ESA model can include various considerations and activities in the solution process and solution management, including: user experience design, technology and tool stack selection, data management, NFR concerns, integration strategies, compliance, team management and communication, risk management, lifecycle management, and so on. ESA requires multidisciplinary knowledge and skills as well as a deep understanding of the solution context and environment.

In other words, A-ESA overlaps with enterprise architecture, solution process, solution management, solution architecture, solution design guidance, and solution modeling. Unlike EA, which is primarily a planning tool, ESA is designed to deliver a working architecture. In addition, ESA is not about sporadic things; it is about a view or an element in the context of a connected architecture. Therefore, modeling is a must.

In practice, a good solution process or solution management framework without a solid architecture model is fragile, unproven, and less useful. This is especially true for a multi-project solution program. ESA or SA without good modeling is at best a conceptual process, method, or framework, not an architecture. Ultimately, an effective A-ESA model clearly reflects an IT architecture with many cross-references from different viewpoints and is an efficacious remedy for the many problems associated with technical debt.

Does Gap Analysis Fall Under ESA?

Sure. Gap analysis in ESA is reflected in the dynamic model and focuses on gap mapping, architectural or technical debt considerations, quality issues, and future solution cases. Metrics mapping, including feasibility assessment and key choice considerations, is an important part of gap analysis, as is the as-is/to-be transition. ESA doesn't include the lengthy assessment or impact analysis that is typically part of the EA deliverable.

Why Is Cost Mentioned in Several Places in this Book?

Architectural thinking inevitably involves cost considerations. The cost factor goes beyond solution architecture and falls under solution management. Many architectural decisions, such as the choice of architectural style, affect cost effectiveness. From an overall cost perspective, it is possible to create a dedicated A-ESA cost profile, which will provide a clear picture of the financial implications.

How Do I Assess A-ESA's Effectiveness and Value?

The effectiveness of A-ESA can be evaluated, in general, from strategic alignment, agility and adaptability, innovation and continuous improvement, business capability enhancement, convergence of technology architecture and business architecture, technology debt and risk management, stakeholder satisfaction, and comparative cost-benefit analysis.

For objective quantitative assessment, you need to define evaluation indicators, collect baseline data and post-implementation data, compare data analysis, establish technology flow evaluation system, and so on.

Bibliography

1. 42010:2011: Systems and Software Engineering – Architecture Description. ISO/IEC/IEEE, 2017.

2. 77 *Building Blocks of Digital Transformation,* An, Jace. Story Tree FDC, 2019.

3. 97 *Things Every Software Architect Should Know,* Monson-Haefel, Richard. O'Reilly, 2009.

4. About the Unified Modeling Language Specification Version 2.5.1. OMG, 2017.

5. *Agile Enterprise Solution Architecture,* Gu, Sean, Vernal Press, 2021.

6. *An Introduction to General Systems Thinking,* Weinberg, G.M. Dorset House, 2001.

7. *Analyzing Requirements and Defining Solution Architectures.* Microsoft Press, 1999.

8. *Architecting Distributed Transactional Applications,* Harrison, Guy. O'Reilly, 2023.

9. *Architecting for Scale,* Atchison, Lee. O'Reilly, 2020.

10. *Architecture Skills Framework,* The Open Group, 2022.

11. *Attributes Guided System Architecture Assessment,* Elias, George, et al. Incose, 2010.

12. *Building Evolutionary Architectures,* Ford, Neal, et al. O'Reilly, 2017.

© Sean (Chunhong) Gu 2024
S. Gu, *Mastering Enterprise Solution Modeling,*
https://doi.org/10.1007/979-8-8688-0992-7

13. *Can Middleware Survive the Serverless enabled Cloud?,* Perera, Srinath. hackernoon.com, 2018.

14. *Capability Management Guide,* WiBotzki, Matthias. Springer Vieweg, 2016.

15. *Clean Architecture,* Martin, Robert. Prentice Hall, 2018.

16. *Cloud Native Architectures,* Laszewski, Tom, et al. Packt, 2018.

17. *Conformance Requirements for the Architect Profession,* The Open Group, 2023.

18. *Critical Questions in Enterprise Architecture Research,* Kotusev, Svyatoslav, et al. "International Journal of Enterprise Information Systems," Volume 13-2, 2017.

19. *Data Service Framework,* COSR. China CITIC Press, 2016.

20. Defining "IT Service" for the IT4IT™ Reference Architecture, The Open Group, 2019.

21. *Digital Doesn't Have to Be Disruptive: The Best Results Can Come From Adaptation Rather Than Reinvention,* Furr, Nathan, et al. "Harvard Business Review," 2019.

22. *Domain Analysis for Microservices. Azure Architecture Center,* 2023.

23. *Domain-Driven Design: Tackling Complexity in the Heart of Software,* Evans, Eric. Addison Wesley, 2003.

24. *Enterprise Solution Architecture: An Overview.* Robertson, Bruce. Gartner Research, 2008.

25. *Everything You Need To Know About Microservices Design,* Kappagantula, Sahiti. edureka.co, 2019.

26. *Fundamentals of Software Architecture,* Richards, Mark, et al. O'Reilly, 2020.

27. *General System Theory: Foundations, Development, Applications,* Ludwig Von Bertalanffy. George Braziller Inc., 2015.

28. *Hands-on Site Reliability Engineering,* Farooqui, Shamayel Mohammed, et al. BPB, 2021.

29. *How to Model an IT Management Landscape and its Transformation using the ArchiMate®, IT4IT™, and TOGAF® Standards,* Van den Brink, Kees, The Open Group, 2019.

30. *Idempotent Consumer Pattern,* Loganathan, Pradeep. Blog, 2023

31. *Implementing Cloud Design Patterns for AWS,* Keery, Sean, et al. Packt Publishing, 2019.

32. *Implementing Domain-Driven Design,* Vernon, Vaughn, et al. Addison-Wesley, 2013

33. *Intelligent Manufacturing Application Interconnection,* GB/T 42405.1-2023, 2023.

34. *Introduction to Solution Architecture,* McSweeney, Alan, ISBN-13:9781797567617, 2019.

35. *IT Governance: How to Reduce Costs and Improve Data Quality Through the Implementation of IT Governance,* Schindlwick, Helmut. CreateSpace, 2017.

36. *Managing Information Technology Projects,* Perera, Srinath, et al. World Scientific, 2023.

37. *Mastering Archimate,* Wierda, Gerben. The Netherlands: R&A, 2014.

38. *Migrating Large-Scale Services to the Cloud,* Passmore, Eric. Apress, 2016.

39. *Model-Based System Architecture,* Weilkiens, Tim, et al. Wiley, 2015.

40. *Next Generation Databases,* Harrison, Guy. Apress, 2015.

41. *Nonfunctional Requirements in Systems Analysis and Design,* Adams, Kevin. Springer, 2015.

42. *On the Criteria to be Used in Decomposing Systems into Modules,* Parnas, S. L. Carnegie Mellon University, 1972.

43. *Serverless Architecture on AWS,* Sbarski, Peter. Manning, 2017.

44. Service Oriented Architecture Modeling Language (SoaML) Specification, OMG, 2012.

45. *Service-oriented Enterprise Application Architecture,* Gu, C. PHEI, 2013.

46. *Software Architecture Metrics,* Ciceri, et al. O'Reilly, 2022.

47. *Software Architecture: The Hard Parts,* Ford, Neal, et al. O'Reilly Media, 2022.

48. *Software Requirements,* Wiegers, Karl, et al. Microsoft Press, 2013.

49. *Solution Architecture Foundations,* Lovatt, Mark. BCS, 2021.

50. *Sustainable Software Architecture,* Lilienthal, Carola. dpunkt.verlag, 2019.

51. *The Agile Architecture Revolution,* Boomberg, Jason. Wiley.

52. *The ArchiMate® 3.2 Specification: an Open Group Standard,* The Open Group, 2022.

53. *The Architecture of Complexity, Herbert A. Simon. Proceedings of the American Philosophical Society,* 1962.

54. *The C4 Model for Visualizing Software Architecture,* Brown, Simon. c4model.com, 2013.

55. *The Cathedral & the Bazaar,* Raymond, Eric. O'Reilly Media, 2001.

56. *The Dark Secret of Enterprise Architecture,* Lewis, Bob. cio.com, 2018.

57. *The Elements of Style,* Strunk, William, et al. Allyn & Bacon, 2000.

58. *The Model Thinker: What You Need to Know to Make Data Work for You,* Page, Scott E. Basic Books, 2019.

59. *The Open Agile Architecture,* The Open Group, 2020.

60. *The Open Group Agile Architecture Framework™,* The Open Group, 2019.

61. *The Science of Lean Software and DevOps,* Forsgren et al, 2018, IT Revolution.

62. *The Timeless Way of Building,* Alexander, Christopher, Oxford University Press, 1979.

Index

A

Abstraction, 178, 181, 387, 390

Accidental architecture, 251

A-ESA architectural process
architecture overview, 173, 174
EA Activity List mapping with A-ESA notations, 175
informal four-step process cycle, 175
mapping between 5W1H method and A-ESA, 176
steps, 173

A-ESA architectural styles, 251
agile (*see* Agile architecture)
application (*see* Application architecture)
architectural evolution, 349, 350
business (*see* Business architecture)
cloud (*see* Cloud architecture)
data (*see* Data architecture)
difference between digital and Agile architecture, 331
digital transformation, 330
EDM (*see* Event-driven microservices (EDM) architecture)

enterprise architecture, 255–257
heat-mapping capability (*see* Heat-mapping capability architecture)
information architecture, 276–278
large website (*see* Large-scale website architecture)
MSA (*see* Microservice architecture (MSA))
RA, 342–344
security, 338–341
SOA, 293
solution, 358
strategic architecture, 253, 254
technical architecture (*see* Technical architecture)

A-ESA Assistive Element List, 452–455

A-ESA base element list, 445–447

A-ESA basic element group, 456

A-ESA Element Usage Scope, 456

A-ESA governance, 412, 413, 434, 436

A-ESA governance techniques
agile modeling, 355
architectural areas, 359
architectural coverage, 363

A-ESA governance
    techniques (*cont.*)
    architectural styles, 358
    architecture processes, 357
    cloud platform environment, 359
    conceptual views, 363
    data information level, 358
    element notations, 363
    elements, 366
    enterprise IT solution, 353
    enterprises, 363
    enterprise solution, 359, 364
    forms, 356
    nine-grid approach, 360
    organizations, 359
    solution operating model, 358
    theme, 354
A-ESA language, 478–479
A-ESA lean model mode
    considerations, 20
    elements, 20, 21
    lean elements notions, 17, 19, 20
    lean view notions, 17, 18
    purpose, 21
A-ESA measurement
    architectural measurement, 197
    common NFRs, 198
    enterprise solutions, 198
    measurement metrics, 200
    NFR authority, 197
    NFRs, 200
    non-functional requirements, 197
    notation mappings, 201
    solution-modeling process, 197

A-ESA model, 421, 458
A-ESA Model Form
    capability model, 355
    correlated views, 355
    governance specifications, 356
    lean model, 355
    profile(s), 356
    sketching, 355
    walkthrough, 355
A-ESA modeling, 456
    architectural clarity, 185
        enhanced architectural
            communication, 189, 190
        unambiguous terms, 186–189
    architectural quality
        cost impact, 192
        enhanced architectural
            conformance, 192–194
    architectural responsibility,
        190, 191
    measurement mappings, 191
A-ESA realization techniques, 386
A-ESA service-based modeling, 192
A-ESA thinking framework, 172
    abstraction and focal
        coverage, 178–180
    known-unknown matrix, 178
    metrics, 184, 185
    model Embodiment, 178
    pillars, 177
    principles, 172
    service-based approach, 172
    system aspect-orientation,
        180–183

Agile approach, 5, 109, 189, 364, 475

Agile architecture, 5

  A-ESA case scenario requirement *vs.* Agile input, 329

  A-ESA mapping with Agile terms, 322–326

  agile transformation, 321

  architectural considerations, 480

  ESA mapping to address issues, 328

  ESA mapping to Agile architecture, 327, 328

  example mapping, 328

  frameworks, 480

  IT architecture, 480

  lifecycle delivery approaches, 321

  reference adoption recommendation, 321, 322

  UX-focused design, 328

Agile enterprise solution architecture (A-ESA), 3, 4, 483

  application governance architecture (*see* Application governance architecture, A-ESA)

  architectural modeling, considerations, 23, 24, 167

  architectural styles, 5

  architecture, 5

  architecture area, 6

  architecture element, 11

    classification, 11–15

    considerations, 15

    derivative elements, 16

    groups, 11

    RQ-NFR, 16

    terms, 16

  architecture view, 7

    considerations, 8

    cross-cutting, 10

    fundamentals, 7, 8

    view definitions, 9, 10

  big data architecture (*see* Big data architecture, A-ESA)

  business architecture (*see* Business architecture)

  CHA (*see* Cloud hosted architecture (CHA)

  chief/lead ESA architect, 26

  customized notations, 26

  EIS architecture (*see* Enterprise integration service (EIS) architecture)

  example solutions profiles, 25

  IT architecture, 167

  IT infrastructure, 151

  metrics, 377

  modeling, 152

  MSA (*see* Microservice architecture (MSA)

  notations, 109, 152, 175, 239, 364

  pattern views, 45

  profiles, 24, 152

  property, 16

  RA (*see* Reference architecture (RA))

  signification service, 397

Agile enterprise solution architecture
(A-ESA) (*cont.*)
single source of truth, 5
specifications, 16, 21
strategic architecture
(*see* Strategic
architecture, A-ESA)
tool, 462
tooling capabilities, 106
Agile environment, 5, 253, 418
Agile ESA model, 7, 24, 355
Agile metrics, 229, 327
Agile model, 452
Agile teams, 412, 413
Agile transformation, 321, 331
AI data considerations, 284–286
AI-driven automation, 444
AI MVI pattern, 348
Airbnb's recommendation
system, 286
AI reference architecture
mapping, 345–347
Application architecture, 384
DevOps environment, 280
functional architecture, 279, 280
IT architectures, 384
microservices, 384
Application domain service, 398
Application governance
architecture, A-ESA, 25, 72
application logic services
(ASs), 88
architecture outline view
(AOV), 73, 74

business logic modules, 72
component domain service
structuring
domains and
dependencies, 92, 93
namespace view, 95, 96
requirements, 96–98
responsibilities, 93–95
component service
realization, 102
considerations, 87
data domain service
determination
characteristics, 99
initial analysis, 99, 100
multidimensional view, 102
NoSQL database, 100, 101
domain partitioning *vs.*
layering, 74
functional services, 85
functional views, 72, 85
generic domain
service, 86, 87
governance function (GF)
element properties, 78–80
elements, 77
role, 81
rules, 81
metric view, 75
element properties, 76
elements, 75–77
expression, 77
order processing
interaction, 91, 92

pattern views, 82
  idempotent consumer
    pattern, 83, 84
  massage-based transaction
    view, 82, 83
  SOA service framework, 84, 85
  TCC transaction view, 82
questions, 159–161
relationship *vs.* interaction, 92
requirements, 86
ShoppingCart
  code snippet property,
    104, 105
  governance property
    specification, 103, 104
  service implementation
    view, 102, 103
SRV-7 service relationship, 88
SRV-7 service relationship
  element properties, 89–91
tabular form *vs.* relationship
  view, 89
tools, 81
user stories, 72, 73
validation views, 106–108
Application logic services (ASs),
  43, 88, 403
Application partitioning practices
  architectural granularity, 396
  decomposition, 393
  MSA Principles, 396
  variations, 394
Application Service
  Adaptability, 212

Application Service
  Maintainability, 212–214
App Logic Service (AS) element,
  182, 279, 290, 291, 294, 296
Architect certification, 468–469
Architectural assessment, 421
  estimation, 422
  levels, 422
Architectural clarity, 185, 189
  enhanced architectural
    communication, 189, 190
  unambiguous terms, 186–189
Architectural connectivity, 207–208
Architectural modeling, 17, 23, 190,
  229, 319, 374
Architectural process, 98, 173–176,
  299–300, 357
Architectural Quality Metrics,
  204, 205
Architectural structuring
  quality, 205–206
Architectural styles, 245, 250
  big data, 249
  cell-based, 248
  clean architecture, 249
  client-server
    architecture, 246
  cloud-based architecture, 247
  ERP architecture, 247
  event architecture, 247
  hexagonal architecture, 246
  mainframe architecture, 246
  mesh architectures, 249
  microkernel architecture, 246

Architectural styles (*cont.*)
  microservices architecture
      (MSA), 248
  mobile app architecture, 247
  modular monolith
      architecture, 248
  monolithic architecture, 247
  multitier architecture, 247
  pipeline architecture, 247
  P2P application
      architecture, 246
  SBA, 248
  shared-nothing
      architecture, 249
  SOA, 248
  supergraph, 249
  UNIX architecture, 246
  web application
      architecture, 246
Architectural theorems, 368
Architectural thinking, 171,
    462, 483
Architectural validation test, 434
Architectural viability, 3, 436
Architectural *vs.* logical
    thinking, 152
Architecture adoption event
    (AAE), 439
  architectural and modeling
      practices, 439
  coverage, 440
  ESA architect, 441
Architecture building block
    (ABB), 4, 255

Architecture decisions, 119, 191,
    227, 274, 319, 379
Architecture model, 16, 34, 141,
    197, 207, 214, 377, 481
Architecture modeling, 377, 417
Architecture pattern, 369
  integration and distributed
      patterns, 369
  pattern frameworks and
      templates, 370
  solution patterns, 370
Architecture Solution Laws, 368
Architecture validation, 360,
    415, 416
Artificial intelligence (AI), 140,
    281, 462
Asset management
  A-ESA models, 437, 438
  elements, 437
Assistive elements, 11, 26, 190, 202,
    364, 457
Atomicity, Consistency, Isolation,
    and Durability (ACID), 99,
    283, 284
Axioms, 367

B

Base elements, 17, 180, 449, 456
Basically Available, Soft-state,
    Eventually consistent
    (BASE), 99
Big data
    analytics, 140, 286

Big data architecture, A-ESA, 25,
140, 249
capabilities, 140, 141
data flow process view, 141, 142
operational view, 148
deployment infrastructure
architecture view, 148, 149
operational capacity
properties, 151
runtime deployment
properties, 149, 150
questions, 165, 166
solution architecture
overview, 142
block view, 142, 143
layer view, 144–146
platform view, 146–148
BizTech-style RAs, 342
Broader Area Scope, 360
Business architecture, 25, 252
and business model, 266–269
business objectives, 273
business scenario, 274, 275
organization communication
structure, 270–272
high-level functional services, 43
IT-based, 35
pattern views, 34, 35
process view mapping, value
stream, 40, 41
capability model, 41
process views, 37–39
vs. BPMN, 39
questions, 154, 155

service interface view, 44
service layering, 42
use case model views, 35–37
Business capability platform, 42
Business decisions
business resiliency, 379
distributed database, 380
identification and rating, 379
performance and recovery, 380
service-level decisions, 380
simulation test, 380
technologies, 380
Business functionality quality, 209
Business Process Execution
Language (BPEL), 39
Business Process Model and
Notation (BPMN), 39,
268, 464

C

Caching architecture, 318, 358
Capacity planning, 96, 215, 320,
404, 422
Cell-based architecture (CBA),
248, 289–292
Clean architecture, 249
Cloud architecture
cloud hosting responsibilities, 331
cloud migration, 333, 334
in A-ESA views, 335
cloud planning mapping with
ESA, 334
cloud service selection, 336, 337

Cloud-based architecture, 125, 247
Cloud design patterns, 139
Cloud hosted architecture
    (CHA), 25, 125
  AWS images, 125, 126
  capabilities, 127, 128
  DevOps view, 133
    deployment packages, 136
    element properties, 135, 136
    maturity level, 136
    properties, 133, 134
    tool chain, 134
  FaaS
    governance view, 130–132
    process view, 128–130
  functional collaboration view, 137
  operational infrastructure view,
    138, 139
  principle *vs.* requirement, 130
  questions, 163, 164
Cloud migration, 125, 333–334
Cloud provider, 332, 333, 336
Coarse-grained systems, 391
Cohesion, 72, 93, 152, 391–392
Command line interface (CLI), 135
Comparative rating techniques
  calculations, 381
  initial analysis, 382
  NFR rating, 381
  tradeoffs, 382
Component-based architecture,
  249, 250
Constraints, 233
  and metrics, 234

Content Distribution Network
  (CDN), 219
Cost, 235
  vs. benefit of deployment
    methods, 433
  considerations, 431
  estimation, 434
  optimization, 434
  vs. sustainability, 432
Cost control and management, 430
CostLin ESA, 235
Cost optimization, 137, 138, 192,
  239, 253, 375
Coupling, 205, 385, 386, 392, 396, 414

D

Data architecture, 281
  AI considerations, 284–286
  common transaction
    patterns, 281
  data transaction considerations,
    282, 283
  and information
    architecture, 281
  transaction considerations,
    283, 284
Data consistency, 120, 123, 283,
  284, 403
Data domain structuring, 402, 403
Data governance, 278, 286, 403
Data mesh architecture, 249
Data service (DS) element, 182
Data services, 43, 88, 215, 407

Data splitting, 318

Decisional architecture
    metrics, 184

Decision thinking, 171

Degrading, 241, 283, 313

Deployment methods, 433

Deployment package (DP)
    element, 183

Deployment Package Mapping, 10,
    176, 295, 328, 404–405

Design thinking, 286, 299, 331, 481

DevOps, 229, 232, 233
    Agile metrics, 229
    architectural issues, 232
    environment, 204, 230, 280, 358
    and KPIs, 231
    metrics, 230

Digital transformation, 299, 330,
    331, 471

Digital Twin Model, 356

Disaster recovery (DR), 165, 227,
    272, 292, 320

Distributed Database Selection,
    378, 380

Distributed database solutions,
    379, 407

Distributed transactional
    architecture, 288–289, 368

Domain-driven design (DDD), 111,
    299, 300, 324, 387

Domain Service Structuring
    Stages, 399

Domain-specific language
    (DSL), 356

Domain Structuring
    Processing, 402

Dynamic application security
    testing (DAST), 163

Dynamic process invocation, 315

E

Economic costs, 430

Edge interface (EG) element, 183

Enterprise architecture (EA), 3,
    255–257, 471
    input architecture, 252
    mappings, 256
    skills, 466

Enterprise business architecture
    (EBA), 259

Enterprise capability, 258, 259,
    262, 263
    purposes, 259

Enterprise Governance
    Measurement, 204

Enterprise integration service (EIS)
    architecture, 25, 52, 53
    business audience, 52
    capability views, 53–55
    custom element images, 52, 53
    deployment view, 71
    element rendering style, 56
    functional service views
        service collaboration
            view, 69, 70
        service collaboration view
            properties, 70

Enterprise integration service (EIS)
architecture (*cont.*)
structural view, 66, 67
structural view
properties, 68, 69
intended audience, 52
metric view, 62, 63
outline views
block view, 60
VFs architecture, 61
view frames (VFs), 61
pattern views, 64, 66
quality requirement tree, 56–58
questions, 157, 158
self-explanatory process
view, 59
use case model view, 58, 59
Enterprise principle
measurements, 202, 203
Enterprise solution architecture
(ESA), 11, 16, 180, 202, 472
Enterprise solution thinking, 173
Enterprise systems
architecture, 472–473
ERP solution architecture, 247
ESA architect role, 5, 465
ESA architectural process, 173
ESA capability element, property
attributes, 260–262
ESA modeling process, 418, 460
ESA solution architect, 5, 373
Event architecture, 247, 311
Event-driven architecture, 125,
172, 318

Event-driven microservices (EDM)
architecture
and microservices, 308
notations, 308–311
Eventual consistency, 101, 117,
283, 400

F

Flexibility *vs.* adaptability, 374
Frequently Asked Questions
(FAQs), 5, 471
Functional architecture, 279,
280, 299
Functional quality metrics, 208–209
Functional requirements, 110,
197, 209
Functional services, 30, 85, 155,
159, 212
Function as a Service (FaaS), 125–127

G

Gap analysis, 175, 274, 449, 482
Generic services (GS), 17, 40, 86
Goodhart's Law, 201
Governance, 412, 442
A-ESA model, 413
architecture, 417
authority, 413
automation, 416
elements, 413
enforcement, 413
function, 414

operational side, 415
role, 417
Granularity, 389–391
concern, 390
Green IT architecture, 238, 363

H

Heat-mapping capability
architecture
capability modeling, 258
enterprise capability, 258, 259,
262, 263
mapping value stream, 263–265
Hexagonal architecture, 246

I, J, K

Information architecture, 34, 254,
276–278, 281
Infrastructure as Code (IaC), 166,
224, 233, 404
IT architecture, 23, 34, 172, 208,
262, 363
IT solution architecture, 185, 242
IT strategic thinking, 171
IT structural thinking, 171

L

Large-scale website
architecture, 312
common rule-based scalability
measures, 312

high-availability architectural
measures, 315–320
rule-based scalability measures,
313, 314
scalability, 312
Leading practices, 25, 192, 203,
333, 367
Load balancing and proxy
service, 316

M

MASA, 249
Measuring requirements
A-ESA, 429
limitations, 428
prioritizing, 429
Mesh architectures, 249
Microkernel architecture, 246
Microservice architecture (MSA),
25, 109, 120, 167, 172, 248,
250, 300–303
aggregate, DDD, 114
aggregates/entities/value
objects, 112
application service rule, 109
architecture decisions, 119
NFRs, 120
architectural process, 299–300
bounded context, 113
component service realization
view, 112
coordination process, 113, 114
data collaboration, 120

Microservice architecture
    (MSA) (*cont.*)
    database considerations,
        121, 122
    data service relationship,
        121, 122
    data service
        representation, 123
    scheduler-agent-supervisor
        pattern, 121
    delivery status element, 114
    design, 301–303
    design views
        bounded context, 111
        business capabilities, 109
        DDD pattern view, 111
        initial case scenario
            view, 110
    DevOps, 303–305
    entity realization relationship,
        112, 113
    functional view, 111
    key considerations, 115
    metrics elements, 115–118
    microservice pattern, 305, 306
    operational environment views,
        123, 124
    operational process, 307
    parts, 109
    polyglot persistence, 123
    principles, 298
    questions, 161, 162
    shapes/colors, 109
    value object *vs.* entity, 114

Microservice design
    application-level, 401
    architectural
        techniques, 401
    decomposability, 401
    single service, 401
    step process, 400
Microservice DevOps, 303–305
Microservice operational
    architecture, 307
Microservice
    pattern, 305, 306
Mid-platform services, 30, 42, 82
Minimal viable architecture model
    (MVAM), 125
Mobile app architecture, 247
Model Adaptation Workshop
    (MAW), 439
Modeling, 40, 171, 194, 201, 319,
    357, 422
Modeling fosters systematic
    thinking, 469
Model view variations, 361
Modularity, 198, 213, 324, 391, 415
Modular monolith, 250
Modular monolith architectural
    style, 248
Monolithic architecture, 247,
    250, 350
MSA, *see* Microservice
    architecture (MSA)
Multi-dimensional
    splitting, 315
Multi-phase transition, 401

Multiple model
modes, 189, 190
Multitier architecture, 247
Mutually exclusive, collectively
exhaustive (MECE)
approach, 152

N

Naming Convention, 16, 460
Network topology, 249
NFR Vague Terms, 373
Nine-grid Architecture Areas, 360
Non-functional requirements
(NFRs), 4, 29, 56, 98, 123,
197, 214, 312, 423
Non-runtime quality, 224, 229, 238

O

Objective quantitative
assessment, 483
Operational architecture, 138, 233,
287, 288, 307, 405
Operational quality, 214–215
Operational quality metrics, 214
Opposing Force Tradeoffs, 236
Order management process system
(OPS), 49
Order management system (OMS),
24, 32, 34, 110
Order Processing Solution
(OPS), 40
Organization design (OD), 272

P

Packaged business capability, 42
Peer-to-peer (P2P) network,
57, 58, 246
Physical architecture, 292
Pipeline architecture, 247
Porter's Five Forces, 266
Process-driven
architecture, 172
Public hosting services, 443

Q

Quality measurements, 425–427
Quality metric indicators, 425

R

Reference architecture (RA), 25,
45, 342–345
architectural style adoption
process, 50
vs. asset, 45
baseline reference
architecture, 50, 51
cell-based, 49
chat AI reference
architecture, 50
commercial level, 45
context-based, 49
data analytics, 47
digital order management, 46
infrastructure view, 48
questions, 156, 157

Remote Procedure Call
(RPC), 84, 350
Requirements mapping, 273,
423, 424
Resource utilization and
allocation, 317
Responsible, Accountable,
Consulted, and Informed
(RACI), 159
Retail business capability
model, 32
Risk management
elements, 436
solution architecture and
management, 437
Robust A-ESA modeling
approach, 444
Robust architecture, 316

S

Security architecture, 166, 338–341
Security attack prevention, 317
Security measurement, 338
Separation, 314
Serverless architecture, 126,
247, 336
Service-based application, 384
abstraction, 387
A-ESA functional services, 387
A-ESA guidance techniques, 385
A-ESA Realization
Techniques, 386
autonomy, 393

cohesion, 391
coupling, 392
granularity, 389
isolation, 393
layering, 388, 389
modularity, 391
separation, 393
Service-based architectural
design, 315
Service-based architecture (SBA),
50, 72, 183, 248, 315
Service interface (SI) element, 182
Service Interface Specification
Techniques, 397
Service level characteristics (SLCs),
4, 92, 109, 182, 229, 379
Service level indicator (SLI), 150,
232, 426
Service-oriented architecture
(SOA), 172, 248, 250, 293
A-ESA mapping, 293
architectural style, 293
mapping to A-ESA, 294, 295
notation mapping between
A-ESA and SOA, 296, 297
Service Realization Template, 370
Shared-nothing architecture, 249
Shared or distributed database, 378
Site Reliability Engineering (SRE),
232, 233
S-MAPS Architectural
Considerations, 239–241
S-MAPS NFR description, 238
S3 model, 4, 5, 173

SOA, *see* Service-oriented architecture (SOA)

Software solution architecture, 464

Software solution model, 356

Solution architecture (SA), 3, 442

Solution Architecture Project Management, 419

Solution building block (SBB), 5, 9, 255, 463

Solution management, 175, 237, 353, 423

Solution management framework, 482

Solution Operating Model (SOM), 180–183, 359

Solution System Model, 180

SOM, *see* Solution Operating Model (SOM)

SonarQube, 134, 230, 417

Sprint approach, 413

Stakeholders, 5, 152, 193, 266, 271, 272, 374, 479

State control, 315

Static application security testing (SAST), 163

Strategic architecture, A-ESA, 25, 27, 253, 254

    capability views, 32, 33

    guiding principles, 27–29

    questions, 153, 154

    solution principles, 29

        AA principle, 30, 31

        BA principle, 30

        TA principle, 31

    subgrouping, 29

Stress test, 216, 320

Sustainability, 5, 167, 237, 239, 372, 431–432

T

Target Operating Model (TOM), 272, 471

Technical architecture, 287, 406

    operational architecture, 287, 288

    technical platform

        CBA mapping, 289–292

        distributed transactional architecture, 288, 289

        physical architecture, 292

Technical architecture techniques

    A-ESA, 404, 405

    application logic, 404

    deployment package mapping, 404

    key attributes, 406

    reliable stateful systems, 408

    self-managing architecture, 406

Technical platform, 288, 289, 406, 411

Tech service (TS) element, 183

Testing process mapping, 435

Test management

    solution management process, 434

Throttle layers, 314

Throttle solutions, 314
Tooling vendors, 77
Tradeoff analysis, 236
    measurement, 375
    solution environment, 377
    steps, 375

U

Unaware-of Requirement
    Mappings, 423
Usability compliance metrics, 210
Use case model (UCM), 10, 36, 209,
    259, 329, 335

User-aware Operational
    Quality, 214–223
User interface (UI) element, 182

V

Value stream, 41, 253, 258,
    263–265, 361
Vertical scaling, 220
Virtual service (VS) element, 183

W, X, Y, Z

Web application architecture, 246